THE MATHEMATICS OF
MAP PROJECTIONS

AN INTRODUCTION TO
THE MATHEMATICS OF
MAP PROJECTIONS

BY

R. K. MELLUISH, M.A.

Late Scholar of Gonville and Caius College, Cambridge. Assistant Master at Rossall School

CAMBRIDGE
AT THE UNIVERSITY PRESS
1931

CAMBRIDGE
UNIVERSITY PRESS

University Printing House, Cambridge CB2 8BS, United Kingdom

Cambridge University Press is part of the University of Cambridge.

It furthers the University's mission by disseminating knowledge in the pursuit of education, learning and research at the highest international levels of excellence.

www.cambridge.org
Information on this title: www.cambridge.org/9781107658486

© Cambridge University Press 1931

First published 1931
First paperback edition 2014

A catalogue record for this publication is available from the British Library

ISBN 978-1-107-65848-6 Paperback

CONTENTS

CONTENTS

PREFACE

This book is the outcome of a mathematical essay on "Maps" written at Cambridge in 1922, and is an attempt to supply the need, which I then discovered, for a book dealing comprehensively with the theories that underlie their construction. It is therefore a book for a student of the mathematical side of geography; and a fair knowledge of the Calculus is all he will need to enable him to fill in for himself the many details of the work which have, for the sake of brevity, been omitted from the text.

I have attempted, in the earlier chapters, first of all to trace, as far as possible, the history of the projections concerned; this is followed by an account of the general theory, from which results are then deduced in the special cases that arise. In this way I deal with the properties of the four main classes of projections. Then follows a chapter on the theory of the Indicatrix, and the method of comparing one projection with another; next the question of finite measurements, perhaps the most important of all from a practical point of view; then a discussion of the best projection for a given country, and finally the general problem of conformal representation of which the complete solution has yet to be effected.

Though I have checked as well as I can the many expansions that are given, I am fully aware that some errors, I hope only slight, may have crept in; and I should therefore be pleased to receive corrections from any reader who may have discovered one.

I am indebted for many valuable suggestions to Mr. A. R. Hinks, F.R.S., Secretary of the Royal Geographical Society, who kindly lent me for reference the works by Germain and Tissot which appear in the list below. A great source of further help, especially in the work on "Minimum Error," was Mr A. E. Young's book, published as No. I of the Technical Series by the same Society.

The following is a complete list of the authors and works consulted, and I trust that, where necessary, adequate acknowledgment to them has been made in the text.

A. GERMAIN. *Traité des Projections des Cartes Géographiques.*
Paris, 1865.

A. TISSOT. *Mémoire sur la Représentation des Surfaces et les Projections des Cartes Géographiques.* Paris, 1881.

J. I. CRAIG. *The Theory of Map Projections.* Cairo, 1910.

A. R. HINKS. *Map Projections.* Cambridge, 1921.

A. E. YOUNG. *Some Investigations in the Theory of Map Projections.* London, 1920.

Encyclopaedia Britannica. Article on Map Projections.

R. K. M

ROSSALL
Oct. 1931

CHAPTER I

INTRODUCTORY

In all probability the drawing of maps and charts was closely associated in its beginnings with the study of geometry. According to Herodotus the Egyptians were brought to the study of geometry in the endeavour to keep records of the extent of their land, so that after the floods of the Nile it might be possible to assess the tax that each man had to pay according to the area of the land left to him. Thus they began to draw diagrams and charts of the divisions of the land, and these, no doubt, eventually grew into maps. Ignorance of the actual shape of the earth probably prevented them from realising many of the difficulties surrounding such a problem as the construction of a map, for ever since those days men have been studying the question and attempting to find the best method of solving it. For the representation of the earth, an oblate spheroid, on a plane, is a problem that admits of no absolutely correct solution. A spheroid is not a developable surface, such as a cone or a cylinder, and thus it is impossible to imagine a piece of paper wrapped round the earth, on which the shapes of countries and continents could be described exactly, and which could then be unwrapped into a plane map.

Any representation that we can make, any map that we can draw, must be incorrect in certain respects. It may be so in all, and be made so that each property of it approximates as nearly as possible to the corresponding one on the earth, or, as is more generally the case, it may be made so as to be correct in one or two respects, and not at all in the others; for example, so as to sacrifice correctness of shape to that of area, or that of azimuth to that of distance.

When a map is being drawn, each point on it is fixed according to some given law which expresses the co-ordinates of that point on the map in terms of those of the corresponding point on the earth. Such a law is called the Projection on which the map is drawn; and the equations of the projection are those which give the relation between the terrestrial co-ordinates and those of the point on the map. It is usually convenient to have the map co-ordinates expanded in ascending powers of the latitude and longitude, or differences of those quantities, of the point under consideration, and these expansions will of course be the solutions of the equations of the projection.

Projections in which the shape of small elements is preserved are called Orthomorphic, those in which areas are preserved Equal Area, and those which give distances of all points from a fixed centre correctly Simple or Equidistant. These are the three chief classes of projections, though there are several others which have been used to a less extent, for example, the Minimum Error, in which the total square error, i.e. the sum of the squares of the errors of scale in two directions at right angles, summed for every point of the map, is made a minimum—and others, such as the doubly azimuthal, of which some account is given later, which have as yet been put to no practical use.

Co-ordinates and length of arc of meridian.

The position of a point on the earth is usually given by its latitude and longitude, but the mathematics is often simplified if, instead of the former of these, we make use of the colatitude, i.e. the angle between the normal at the point and the polar axis.

Let P be any point on the earth, PG the normal at P, and NOG the axis of the earth; and let $P\hat{G}N = \theta$, and the longitude of the meridian NPA be ϕ. Then if we take the earth as having an equatorial radius unity and eccentricity ϵ, the equation of the meridian NPA referred to its principal axes is

$$x^2 + \frac{y^2}{1 - \epsilon^2} = 1.$$

It will sometimes be found convenient to use, instead of the eccentricity ϵ, the ellipticity, i.e. the ratio of the difference between the semi-axes to the semi-major axis. Calling this e we have

$$(1 - e)^2 = 1 - \epsilon^2,$$

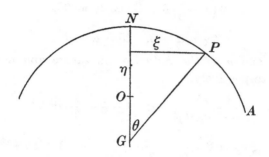

or, neglecting powers of e above the first, $e = \dfrac{\epsilon^2}{2}$. Thus the equation of the meridian is

$$x^2 + \frac{y^2}{1 - 2e} = 1.$$

If P be the point ξ, η the equation of the normal PG is

$$\frac{x - \xi}{\xi} = \frac{y - \eta}{\eta} (1 - \epsilon^2),$$

whence

$$\tan \theta = \frac{\xi (1 - \epsilon^2)}{\eta}.$$

Also

$$\xi^2 + \frac{\eta^2}{1 - \epsilon^2} = 1,$$

$$\therefore \quad \xi = \frac{\sin \theta}{(1 - \epsilon^2 \cos^2 \theta)^{\frac{1}{2}}} \quad \text{or} \quad \sin \theta (1 + e \cos^2 \theta),$$

$$\eta = \frac{(1 - \epsilon^2) \cos \theta}{(1 - \epsilon^2 \cos^2 \theta)^{\frac{1}{2}}} \quad \text{or} \quad \cos \theta \{1 - e (2 - \cos^2 \theta)\}.$$

To find the element $d\sigma$ of arc of the meridian we have

$$\frac{d\sigma}{d\theta} = \left[\left(\frac{d\xi}{d\theta}\right)^2 + \left(\frac{d\eta}{d\theta}\right)^2\right]^{\frac{1}{2}} = \frac{1-\epsilon^2}{(1-\epsilon^2\cos^2\theta)^{\frac{3}{2}}},$$

or $\quad\quad\quad\quad\quad 1 - \dfrac{e}{2} + \dfrac{3e}{2}\cos 2\theta;$

and the length of the arc of a meridian between two points of colatitudes α and β is to the first order

$$2\left(1 - \frac{e}{2}\right)\delta + \frac{3e}{2}\cos 2\chi \sin 2\delta,$$

where $\quad\quad\quad 2\chi = \alpha + \beta, \quad 2\delta = \alpha - \beta.$

Also if ρ and ν be the radius of curvature and normal of meridian respectively,

$$\rho = \frac{1-\epsilon^2}{(1-\epsilon^2\cos^2\theta)^{\frac{3}{2}}} \quad \text{or} \quad 1 - \frac{e}{2} + \frac{3e}{2}\cos 2\theta,$$

$$\nu = \xi\csc\theta = \frac{1}{(1-\epsilon^2\cos^2\theta)^{\frac{1}{2}}} \quad \text{or} \quad 1 + e\cos^2\theta.$$

Normal, oblique and transverse projections.

If we neglect e altogether and regard the earth as a sphere, which is quite often sufficient for maps on small scales such as are used in atlases, then we simply have

$$\xi = \sin\theta, \quad \eta = \cos\theta.$$

In this case, since all the meridians are circles, it becomes possible to measure the latitude and longitude from an axis other than the geographical one. Thus out of one projection, giving the co-ordinates of the map point in terms of the geographical latitude and longitude of the earth point, we may derive another by regarding these latter as being no longer referred to the polar axis but to an axis through some other point. The original projection is called Normal, and the derived one Oblique, or, if the pole of the map be on the Equator, Transverse.

To obtain the relations between geographical co-ordinates and those referred to another pole, when regarding the earth as a sphere, we make use of the ordinary results of spherical trigonometry.

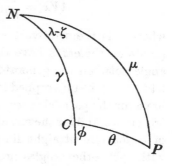

Let C be the pole of the map, N the north pole, and the co-latitude and longitude of C be γ, ζ. Then if P is any point whose co-ordinates referred to N are μ, λ and referred to C are θ, ϕ, we have—since, as we may measure ϕ from any plane, we may suppose it measured from that of the meridian through C—

$$\cos \theta = \cos \mu \cos \gamma + \sin \mu \sin \gamma \cos (\lambda - \zeta),$$

$$\cos \mu = \cos \theta \cos \gamma - \sin \theta \sin \gamma \cos \phi,$$

$$\sin \theta \sin \phi = \sin \mu \sin (\lambda - \zeta),$$

from which we may derive, by eliminating μ,

$$\cos \theta \sin \gamma + \sin \theta \cos \gamma \cos \phi = \sin \theta \sin \phi \cot (\lambda - \zeta).$$

These equations enable us, when given the co-ordinates of a point on the map in terms of θ and ϕ, to find equations between them and their geographical co-ordinates μ and λ.

Conical projections.

A conical projection is one in which the plane map is derived from one drawn on a cone, simply unwrapping it from the cone. Imagine a cone, vertex V, placed so that its axis coincides with that of the earth, as in the figure, and suppose the projection be made in such a way that the azimuths of all points are preserved and the parallels on the earth become the circular cross-sections of the cone. Then we shall have a map on the cone in which the meridians on the earth are represented by the generators. Now if P be a point, longitude ϕ, and A the point on the meridian $0°$ and of the same latitude

$$A\hat{V}P = \frac{AP}{VP} = \frac{CP}{VP}\,\phi = \phi\sin\alpha,$$

where α is the semi-vertical angle of the cone. Thus if two points on the earth have their longitudes differing by ϕ, the angle between the generators through them is $\phi\sin\alpha$. Now let the map be unwrapped from the cone into a plane; we shall have for the parallels arcs of concentric circles, and for the meridians a set
of concurrent straight lines making
with each other angles proportional
to the differences in longitude, and
the constant of the proportion, which
is called the constant of the cone, is
sin α. Thus a map of the world would
be enclosed between two lines inclined
at an angle of $2\pi\sin\alpha$.

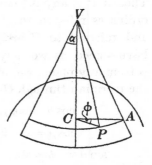

In the particular cases when the semi-vertical angle of the cone has its limiting values 0 and $\pi/2$, the cone becomes in the first case a cylinder and the projection is said to be cylindrical, in the second a plane and the projection is called Azimuthal, since the azimuths of all points from the pole are given correctly. A projection of this latter kind has also been called Zenithal, but there seems to be little reason for such a name.

A conical, cylindrical or azimuthal projection can of course be made to satisfy any other condition, for we have still to specify in what way the map on the cone is derived from the earth; for example it may be derived so that shapes are pre-served, giving an orthomorphic, or areas, giving an equal area projection. Again the cone may be applied, if we regard the earth as a sphere, with its axis coinciding with some diameter other than the polar axis, and in this case we should have an oblique conical projection.

The equations eventually obtained by these considerations will contain certain constants, e.g. the constant of the cone, and we may determine these so as to satisfy certain other kinds of conditions. One method is to make the lengths of two parallels correct, that is, equal to the lengths of the corresponding parallels on the earth, subject to the modification of the scale on which the map is drawn. The cone of the fig. on p. 6, or the cylinder, in the case of a cylindrical projection, cuts the earth in these two parallels, which are said to be Standard Parallels. A modification of this method is to make the cone or cylinder touch the earth, in which case the two parallels coincide and the projection has one standard parallel only.

In the case of a conical projection with two standard parallels of colatitudes θ_1 and θ_2, and radii on the map r_1 and r_2, we shall have, taking the radius of the earth as unity— as we shall do throughout—

$$2\pi \sin \theta_1 = 2n\pi r_1,$$

and therefore $\quad \sin \theta_1 = nr_1$ or $\xi_1 = nr_1,$

$$\sin \theta_2 = nr_2, \quad \xi_2 = nr_2$$

for the spheroid. From these two equations n and the other constant which appears in the expression for the radius may be found.

Another method of calculating the constants, which has been used by Sir George Airy, Col. Clarke, and more latterly by Mr A. E. Young, is that of Minimum Error, a method practicable only in the case of conical projections. Suppose P any point on the map, co-ordinates θ, ϕ, and let the parallel through it have a radius r. Let Q be a neighbouring point on the same meridian, co-ordinates $\theta + \delta\theta$, ϕ; and R a neighbouring point on the same parallel,

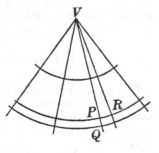

co-ordinates θ, $\phi + \delta\phi$. Then the angle subtended at the centre V by PR is $nd\phi$, where n is the constant of the cone. We thus have $PQ = \delta r$, $PR = nr\,\delta\phi$; and the corresponding lengths on the earth are $\delta\theta$ and $\sin\theta\,\delta\phi$. Thus the scales along the meridian and the parallel are respectively

$$\frac{dr}{d\theta} \quad \text{and} \quad \frac{nr}{\sin\theta}.$$

Sir George Airy devised the plane of making the total square error, i.e. the sum of the squares of the errors of scale in these two directions, summed for every point of the map, a minimum. We should thus have to make

$$\iint \left[\left(\frac{dr}{d\theta} - 1\right)^2 + \left(\frac{nr}{\sin\theta} - 1\right)^2 \right] \sin\theta\,d\theta\,d\phi$$

a minimum, where the integration is taken for the whole area of the map, the factor $\sin\theta\,d\theta\,d\phi$ being the element of area on the earth, regarded as a sphere*. Airy only considered the case $n = 1$, i.e. an azimuthal projection, and found an expression for r in terms of θ, but Young, in his *Some Investigations in the Theory of Map Projections*, extended the method to the calculation of the constants of the projections as well as to the discovery of the equations.

A third method of calculating the constants, probably introduced first by Murdoch, is to make the total area of the map correct. As an example take the conical projection of the zone between two parallels α and β, the radii of whose projections are r_α and r_β. The area on the map is $(r_\beta{}^2 - r_\alpha{}^2)\,n\pi$, and that on the earth is

$$\int_\alpha^\beta \int_0^{2\pi} \sin\theta\,d\theta\,d\phi = 2\pi(\cos\alpha - \cos\beta).$$

---

* In a conical projection r is independent of ϕ, and if, as is usual, we are calculating the total square error of a zone between two parallels α and β, we can integrate with respect to ϕ at once, between the limits 0 and 2π, and dividing by 2π obtain the expression which we call the total square error

$$M = \int_\alpha^\beta \left[\left(\frac{dr}{d\theta} - 1\right)^2 + \left(\frac{nr}{\sin\theta} - 1\right)^2 \right] \sin\theta\,d\theta.$$

Thus we have to satisfy the equation

$$n\,(r_\beta{}^2 - r_a{}^2) = 2\,(\cos\alpha - \cos\beta).$$

Because in this chapter we have spoken of conical projections, including azimuthal and cylindrical, and introduced methods applicable to them, and sometimes to them alone, it must not be imagined that there are no other kinds. We have mentioned them first because historically they come first, and others have been derived from them, or have been found, after their discovery, to be a transverse or oblique application of one or other of them.

In the following chapters we may say that we have this problem before us: first to find equations that will give a map of the earth, regarded as a sphere or a spheroid, having one or more properties correct; secondly to fix the constants of those equations in such a way that the inaccuracies in the other properties shall be reduced to a minimum; and thirdly to find out how, supposing the map before us, we are to use it to measure scales, distances, angles and bearings from one point to another.

CHAPTER II

THE SIMPLE CONIC

According to Tissot the earliest known maps were constructed by the Egyptians as early as the reign of Sesostris, 1600 B.C., though there is no definite mention of any particular map until 545 B.C., when Anaximander of Milet, a disciple of Thales, is said to have constructed one, but on what projection we do not know. About 300 B.C. Dicearcus, a disciple of Aristotle, produced a new map, and though we do not know this for certain yet it is probable that the projection was what is known as the Carte Plate, or Simple Cylindrical with one standard parallel. Certainly it was on this projection that the map of Eratosthenes was constructed, and also those of all the Greeks and Romans until Ptolemy (130 B.C.), who used it for maps of countries, but for one of the world—180° in longitude and a maximum of 80° in latitude—the simple conical projection. We are thus led to begin by studying this latter projection, of which the former is a particular case.

The earth is projected first on to a cone in such a way that distances along the meridians are preserved; the cone is then unwrapped into a plane, and the meridians, which were the generators of the cone, become inclined to one another at angles proportional to the difference of their longitudes. Let us first of all neglect the ellipticity of the earth; then since our meridian scale must be correct we have $\dfrac{dr}{d\theta} = 1$ and hence $r = a + \theta$, where a is a constant that has to be determined. Having found r, we have for the Cartesian co-ordinates of any point, referred to the central meridian as axis of y,

$$x = r \sin n\phi,$$

$$y = r \cos n\phi,$$

where n is the constant of the cone.

Standard parallels.

Suppose we choose our two standard parallels θ_1 and θ_2; and let
$$\theta_1 + \theta_2 = 2\chi, \quad \theta_2 - \theta_1 = 2\delta.$$

Then
$$n\,(a + \theta_1) = \sin\theta_1,$$
$$n\,(a + \theta_2) = \sin\theta_2,$$

and we have
$$n = \cos\chi\,\frac{\sin\delta}{\delta} = \cos\chi\left(1 - \frac{\delta^2}{6} + \frac{\delta^4}{120}\cdots\right),$$

and
$$a + \chi = \tan\chi\delta\cot\delta = \tan\chi\left(1 - \frac{\delta^2}{3} - \frac{\delta^4}{45}\cdots\right).$$

Thus the radius of the parallel θ is
$$r = \theta - \chi + \tan\chi\delta\cot\delta$$
$$= -l + \delta\tan\chi\cot\delta,$$

where l is the difference of latitude measured, positive when northward, from the mean parallel χ. Thus taking our axis of x at right angles to the mean meridian at the intersection with it of the mean parallel
$$x = r\sin n\phi, \quad y = \delta\tan\chi\cot\delta - r\cos n\phi.$$

$$\therefore\quad x = \phi\sin\chi\cos\delta - l\phi\cos\chi\,\frac{\sin\delta}{\delta} - \phi^3\frac{\cos^2\chi\sin\chi\sin^2\delta\cos\delta}{6\delta^2}$$
$$+\, l\phi^3\frac{\cos^3\chi\sin^3\delta}{6\delta^3}\cdots,$$

$$y = l + \frac{\sin 2\chi\sin 2\delta}{8\delta}\phi^2 - \frac{\cos^2\chi\sin^2\delta}{2\delta^2}l\phi^2$$
$$-\, \frac{\sin\chi\cos^3\chi\sin^3\delta\cos\delta}{24\delta^3}\phi^4\cdots.$$

We see at once that if we wish to obtain a projection with one standard parallel, i.e. when the cone of projection touches the sphere, we must make $\delta = 0$. This gives $n = \cos\chi$ and
$$x = \phi\sin\chi - l\phi\cos\chi - \cos^2\chi\sin\chi\frac{\phi^3}{6} + \cos^3\chi\frac{l\phi^3}{6}\cdots,$$

$$y = l + \tfrac{1}{2}\sin\chi\cos\chi\phi^2 - \cos^2\chi\frac{l\phi^2}{2} - \sin\chi\cos^3\chi\frac{\phi^4}{24}\cdots.$$

In this projection the scale along the meridians is everywhere correct, but along the parallels it will only be so on the two (or one) standard ones. At the value $\theta = \chi$ we see that $nr = \sin\chi\cos\delta$, and thus the scale along the mean parallel, which is $\dfrac{nr_\chi}{\sin\chi} = \cos\delta$, is < 1. Hence the scale between the standard parallels is less than unity and for values of $\theta > \theta_2$ or $< \theta_1$ it will be greater than unity. Also the scale of areas will vary in the same way seeing that the meridian scale is correct.

Up to the present we have been regarding the earth as a sphere, which will not give sufficient results for maps on large scales; if, then, we introduce the ellipticity we have to write, instead of θ,

$$\sigma = \int \frac{(1-\epsilon^2)\,d\theta}{(1-\epsilon^2\cos^2\theta)^{\frac{3}{2}}} = \left(1 - \frac{e}{2}\right)\theta + \frac{3e}{4}\sin 2\theta,$$

neglecting higher powers of e. The distance $2L$ along the meridians between the standard parallels is given by

$$L = \left(1 - \frac{e}{2}\right)\delta + \frac{3e}{4}\cos 2\chi \sin 2\delta,$$

and

$$n = \frac{\cos\chi\sin\delta}{L}\{1 + e\,(1 - 3\sin^2\chi\cos^2\delta - \cos^2\chi\sin^2\delta)\}.$$

The radius of the mid-parallel

$$r_\chi = L\tan\chi\cot\delta\,\{1 + e\,(\cos 2\delta - \cos 2\chi)\} + \frac{3e}{2}\sin 2\chi\sin^2\delta,$$

and with the same axes as before we have for the co-ordinates

$$x = (r_\chi - M)\left(n\phi - \frac{n^3\phi^3}{6} + \dots\right),$$

$$y = M - (r_\chi - M)\left(\frac{n^2\phi^2}{2} - \frac{n^4\phi^4}{24} + \dots\right),$$

where M is the meridian distance

$$\left(1 - \frac{e}{2}\right)(\chi - \theta) + \frac{3e}{4}(\sin 2\chi - \sin 2\theta),$$

measured from the mid-parallel,

$$= \left(1 - \frac{e}{2} + \frac{3e}{2}\cos 2\chi\right) l + 3e\sin\chi\cos\chi l^2 \ldots.$$

The particular cases of this projection found by making n respectively 1 and 0 are the simple azimuthal, or the Azimuthal Equidistant, and the Simple Cylindrical. In the former case the cone of projection becomes a plane and we have one standard parallel only: thus $\cos\chi = 1$, $\chi = 0$, the standard parallel is the pole and the radius of any parallel

$$\left(1 - \frac{e}{2}\right)\theta + \frac{3e}{4}\sin 2\theta.$$

This projection is used only for maps of the polar regions though it was proposed by Guillaume Postel in 1581 for a map of the world bounded by the equator.

In the latter case $n = 0$, and $\cos\chi\,\dfrac{\sin\delta}{L} = 0$, thus $\chi = \dfrac{\pi}{2}$ and the standard parallels are equidistant from the equator

$$x = \phi\cos\delta, \quad y = M = \left(1 - \frac{e}{2}\right)\left(\frac{\pi}{2} - \theta\right) - \frac{3e}{4}\sin 2\theta.$$

The graticule, i.e. the two sets of lines formed by the parallels and meridians on the map, are thus two sets of parallel lines at right angles; and the projection has therefore been called the Plate Parallelogrammatique. In the more common case of the Plate Carrée there is only one standard parallel, the equator, and since $\delta = 0$ we have for the co-ordinates

$$x = \phi, \quad y = M.$$

This is the projection which, as we have already noted, was probably the first used of all for any map. After the time of Ptolemy it began to be superseded by the Simple Conic—which was used by Mercator for a map of Europe in 1554, and by de L'Isle in 1745 for one of Russia—and by a variation of it which we will consider in the next chapter—and when in 1569 Mercator produced his nautical chart it

went out of use almost entirely, though D'Auville used it in
1776 for a map of Guinea, since it was suitable for a country
of small extent of latitude in the neighbourhood of the
equator. At about the same time François Cassini and his
son Dominique were working on a map of France, begun in
1745 but not completed for forty-eight years, on the transverse
plate carrée projection. Disregarding the ellipticity the normal
projection is

$$x = \phi, \quad y = \frac{\pi}{2} - \theta.$$

Substituting in the equations of p. 5 we have for the parallels
of this transverse projection

$$\cos y \cos x = - \cos \chi,$$

and for the meridians

$$\sin x = \tan y \tan (\lambda - \zeta).$$

Solving for x and y we obtain

$$\tan x = - \tan \mu \sin (\lambda - \zeta),$$
$$\sin y = \sin \mu \cos (\lambda - \zeta).$$

It will readily be seen that if N is the north pole, P any point,
and PM a great circle at right angles to the meridian NM,
of longitude $\pi/2 + \zeta$, that x and y are proportional to the arcs
NM and PM respectively. It was in this way that Cassini
obtained his projection, and not by considering it as the
transverse plate carrée. If we supposed the longitude measured
from the central meridian we should have

$$x = \tan^{-1} (\tan \mu \cos D) = \mu - \tfrac{1}{2}D^2 \cos \mu \sin \mu$$
$$- \tfrac{1}{24}D^4 \sin \mu \cos \mu (5 - 6 \cos^2 \mu),$$
$$y = \sin^{-1} (\sin \mu \sin D) = D \sin \mu - \tfrac{1}{6}D^3 \sin \mu \cos^2 \mu$$
$$+ \tfrac{1}{120}D^5 \sin \mu (1 - 10 \sin^2 \mu + 9 \sin^4 \mu),$$

or in terms of the latitude l, and reversing the direction of x,

$$x = l + \tfrac{1}{2}D^2 \cos l \sin l + \tfrac{1}{24}D^4 \sin l \cos l (5 - 6 \sin^2 l),$$
$$y = D \cos l - \tfrac{1}{6}D^3 \cos l \sin^2 l + \tfrac{1}{120}D^5 \cos l (1 - 10 \cos^2 l + 9 \cos^4 l).$$

Let it be noted that as this is a transverse projection x and y have become interchanged.

Less well known than this is the transverse simple azimuthal, studied in 1772 by Lambert and in 1799 by Cagnoli. We will consider it as a special case of the oblique simple azimuthal, which appears to be of more value, as the centre of the map, from which bearings and distances of all points are given truly, is not restricted to the equator but may be some important position such as London or New York.

The equations of the normal projection are

$$x = \theta \sin \phi, \quad y = \theta \cos \phi,$$

and if we use polar co-ordinates r, ω, we have

$$\theta = r, \quad \phi = \frac{\pi}{2} - \omega.$$

Thus the equation of the parallel on the map is

$$\cos \mu - \cos r \cos \gamma = \sin r \sin \gamma \sin \omega,$$

where γ, ζ is the centre of the map. Squaring both sides and rearranging, we obtain

$$\operatorname{cosec}^2 \gamma \, (\cos r - \cos \mu \cos \gamma)^2 + \sin^2 r \cos^2 \omega = \sin^2 \mu;$$

for the poles, $\mu = 0$ gives $\cos r = \cos \gamma$ and $\cos \omega = 0$, and $\mu = \pi$ gives $\cos r = - \cos \gamma$, $\cos \omega = 0$. Thus the poles are the points 0, γ; 0, $- (\pi - \gamma)$. The equation of the equator is

$$\cot r + \tan \gamma \sin \omega = 0.$$

For the meridians we have

$$\cos r + \sin r \cot \gamma \sin \omega = \sin r \cos \omega \operatorname{cosec} \gamma \cot (\lambda - \zeta).$$

When $\lambda - \zeta = 0$ or π we have

$$\sin r \cos \omega = 0,$$

which gives $\qquad r = \pi \ \text{ or } \ \omega = \dfrac{\pi}{2}.$

Thus the central meridian is the line $\omega = \dfrac{\pi}{2}$, i.e. the axis of y; and the boundary of the map is the circle $r = \pi$.

It can be shewn by actual drawing that the parallels, except those whose colatitudes are greater than $\pi - \gamma$, are all oval curves surrounding the north pole; those with greater colatitudes surround the south pole; and the meridians are curves passing through the poles and, in the region near the centre of the map, cutting the parallels nearly at right angles. Those near the boundary have sharp bends as they near colatitude $\pi - \gamma$, which produce great distortion in shape in the region of the antipodes.

We have, from the original equations of the last chapter,

$$\cos r = \cos \mu \cos \gamma + \sin \mu \sin \gamma \cos D,$$

where $D = \lambda - \zeta$. In expanding the co-ordinates it will be found best to expand first in powers of D and afterwards in powers of l, the difference of latitude $\gamma - \mu$. We have

$$x = r \cos \omega = r \sin \phi = \frac{r \sin (\gamma - l) \sin D}{\sin r},$$

$$y = r \sin \omega = r \cos \phi = \frac{- r \{\cos (\gamma - l) - \cos r \cos \gamma\}}{\sin r \sin \gamma},$$

and expanding by Maclaurin's theorem and reversing the direction of y, we find

$$x = D \sin \gamma - l D \cos \gamma - \frac{l^2 D}{3} \sin \gamma - \frac{D^3}{6} \sin \gamma \cos^2 \gamma$$
$$+ \frac{l D^3}{6} \cos \gamma \cos 2\gamma,$$

$$y = l + \frac{D^2}{2} \sin \gamma \cos \gamma - \frac{l D^2}{6} (3 - 4 \sin^2 \gamma) - \frac{l^2 D^2}{12} \sin \gamma \cos \gamma$$
$$- \frac{D^4}{96} \sin 4\gamma.$$

Note that these agree, as far as terms of the second degree, with the expansions of the normal simple conic with one standard parallel of colatitude γ.

We now come to the consideration of the general projection with meridian scale true. This projection, which obviously includes the simple conic, has been used in one of its forms by Euler for his map of Russia with two standard parallels and by Murdoch, the rector of Stradishall in Suffolk, who described his two projections in a paper published in the *Philosophical Transactions of the Royal Society* in 1758. The following investigation of the complete projection is due to Young. For the meridian scale to be true we must have $\frac{dr}{d\theta} = 1$, where r is the radius of a parallel. Hence $r = a + \theta$, a being a constant which may be determined in many different ways. For his map of Russia, lying in the zone between two parallels, Euler chose a so as to make the absolute errors on the extreme parallels equal to one another, and opposite to that on the mid-parallel; there will be two standard parallels, not chosen beforehand but depending on the size and position of the zone.

Suppose the constant of the cone n, and the extreme parallels α, β, and as before let $\alpha + \beta = 2\chi$, $\alpha - \beta = 2\delta$. Then the condition gives

$$2\pi \{n (a + \chi - \delta) - \sin (\chi - \delta)\} = 2\pi \{\sin \chi - n (a + \chi)\}$$
$$= 2\pi \{n (a + \chi + \delta) - \sin (\chi + \delta)\},$$

whence $\quad n = \dfrac{\cos \chi \sin \delta}{\delta} = \cos \chi \left(1 - \dfrac{\delta^2}{6} + \dfrac{\delta^4}{120} \cdots \right),$

and $\quad a + \chi = \dfrac{\delta}{2} \tan \chi \cot \dfrac{\delta}{2} = \tan \chi \left(1 - \dfrac{\delta^2}{12} - \dfrac{\delta^4}{720} \cdots \right).$

If instead of using the total errors we take the errors of scale, we have the conditions

$$\frac{n (a + \chi - \delta)}{\sin (\chi - \delta)} - 1 = 1 - \frac{n (a + \chi)}{\sin \chi} = \frac{n (a + \chi + \delta)}{\sin (\chi + \delta)} - 1,$$

giving $\quad n = \dfrac{2}{\delta} \tan \dfrac{\delta}{2} \cos \chi = \cos \chi \left(1 + \dfrac{\delta^2}{12} - \dfrac{\delta^4}{120} \cdots \right),$

and $\quad a + \chi = \delta \tan \chi \cot \delta = \tan \chi \left(1 - \dfrac{\delta^2}{3} - \dfrac{\delta^4}{45} \cdots \right).$

Murdoch, in his first projection, takes $n = \cos \chi$, as in the simple conic with one standard parallel, and also makes the total area of the map true. This gives

$$\cos \alpha - \cos \beta = n \{(a + \beta)^2 - (a + \alpha)^2\},$$

i.e. $$a + \chi = \tan \chi \frac{\sin \delta}{\delta} = \tan \chi \left(1 - \frac{\delta^2}{6} + \frac{\delta^4}{120} \cdots \right):$$

In his third projection, which is considered by Young to be the most elegant of all conical projections, he takes

$$a + \chi = \delta \tan \chi \cot \delta,$$

as in the last but one, and chooses n so as to make the total area true. This gives

$$n = \cos \chi \frac{\sin^2 \delta}{\delta^2 \cos \delta} = \cos \chi \left(1 + \frac{\delta^2}{6} + \frac{31\delta^4}{360} \cdots \right).$$

It will be seen that in this projection the scales along the extreme parallels are each $\dfrac{\tan \delta}{\delta}$, but the errors are no longer equal and opposite to that on the mean parallel which becomes

$$1 - \frac{\sin \delta}{\delta}.$$

In order to compare these projections with each other and the simple conic with two standard parallels, having regard to general efficiency, let us calculate the value of the total square error M. We have

$$M = \pm \int_\alpha^\beta \left(\frac{n\,(a + \theta)}{\sin \theta} - 1 \right)^2 \sin \theta \, d\theta$$

$$= \pm [\cos \alpha - \cos \beta - 4na\delta - 4n\chi\delta + n^2 a^2 S_0 + 2n^2 a S_1 + n^2 S_2],$$

where $$S_0 = \int_\alpha^\beta \frac{d\theta}{\sin \theta}, \quad S_1 = \int_\alpha^\beta \frac{\theta \, d\theta}{\sin \theta}, \quad S_2 = \int_\alpha^\beta \frac{\theta^2 d\theta}{\sin \theta}.$$

To obtain these functions we make use of the general expansion of an integral in terms of the mean value and difference between the limits, viz.

$$\int_\alpha^\beta f(\theta) \, d\theta = 2\delta f(\chi) + \frac{2\delta^3}{3!} f''(\chi) + \frac{2\delta^5}{5!} f^{iv}(\chi) \cdots,$$

which can be obtained from the general expansion of any function by Taylor's theorem. Suppose

$$S_0 = 2\delta \left(A_0 + A_2 \frac{\delta^2}{3!} + A_4 \frac{\delta^4}{5!} \cdots \right),$$

$$S_1 = 2\delta \left(B_0 + B_2 \frac{\delta^2}{3!} + B_4 \frac{\delta^4}{5!} \cdots \right),$$

$$S_2 = 2\delta \left(C_0 + C_2 \frac{\delta^2}{3!} + C_4 \frac{\delta^4}{5!} \cdots \right).$$

Then

$$A_r = \frac{d^r}{d\chi^r} (\operatorname{cosec} \chi),$$

$$B_r = \frac{d^r}{d\chi^r} (\chi \operatorname{cosec} \chi) = \chi A_r + r A_{r-1},$$

$$C_r = \frac{d^r}{d\chi^r} (\chi^2 \operatorname{cosec} \chi) = \chi^2 A_r + 2r\chi A_{r-1} + r(r-1) A_{r-2},$$

by Liebnitz' theorem.

Substituting for the B's and C's we obtain

$$M = \pm \left\{ 2 \sin \chi \sin \delta - 4n\delta (a + \chi) \right.$$
$$+ 2n^2 \delta \left[(a + \chi)^2 A_0 + \{(a + \chi)^2 A_2 + 2(a + \chi) A_1 + 2A_0\} \frac{\delta^2}{3!} \right.$$
$$+ \{(a + \chi)^2 A_4 + 8(a + \chi) A_3 + 12A_2\} \frac{\delta^4}{5!} \cdots$$
$$+ \{(a + \chi)^2 A_{2r} + 4r(a + \chi) A_{2r-1}$$
$$\left. \left. + 2r(2r - 1) A_{2r-2}\} \frac{\delta^{2r}}{(2r + 1)!} \cdots \right] \right\}.$$

Now if we suppose

$$a + \chi = \tan \chi (1 + p\delta^2 + p'\delta^4)$$

and

$$n = \cos \chi (1 + q\delta^2 + q'\delta^4),$$

we have always, as far as terms in δ^5,

$$M = \frac{\delta^5}{30} \sin \chi \{60 (p + q)^2 + 20 (p + q) + 3\}.$$

If we calculate this for the simple conical projection of a zone bounded by its standard parallels α, β, we find

$$M = \frac{\delta^5}{10} \sin \chi.$$

For Euler's map of Russia, and the allied projection in which the errors of scale on the extreme parallels are equal,

$$M = \frac{7}{120} \delta^5 \sin \chi,$$

and for Murdoch's first and third projections,

$$M = \frac{2}{45} \delta^5 \sin \chi.$$

This last value of M satisfies the condition for minimum error in this type of projection—and, as we shall see later, for all conical projections—for if we differentiate M with respect to a and n, and put $\dfrac{dM}{da}$ and $\dfrac{dM}{dn}$ equal to zero in turn, we obtain

$$a = \frac{2\delta - nS_1}{nS_0} \quad \text{and} \quad n = \frac{2\delta (a + \chi)}{a^2 S_0 + 2aS_1 + S_2} = \frac{2\delta (\chi S_0 - S_1)}{S_0 S_2 - S_1^2},$$

and $\qquad M = \pm \{2 \sin \chi \sin \delta - 2n\delta (a + \chi)\},$

giving $\quad n = \cos \chi \left(1 + \dfrac{\delta^2}{30}\right), \quad a + \chi = \tan \chi \left(1 - \dfrac{\delta^2}{5}\right),$

and $\qquad\qquad M = \dfrac{2\delta^5}{45} \sin \chi.$

It will be convenient at this point, for purposes of comparison, to tabulate (see p. 21) the six simple projections that we have been considering, so that the differences in the expressions for the co-ordinates may be noted.

Now let us consider the simple cylindrical projections in the same way. By definition n must vanish, and therefore $\chi = \dfrac{\pi}{2}$, and we must have always $y = \dfrac{\pi}{2} - \theta$.

In the case of the plate carrée we had $x = \phi$, therefore we

Projection	n	$a + \chi$	Co-ordinates
Simple conic with two standard parallels	$\cos\chi\,\dfrac{\sin\delta}{\delta}$ $= \cos\chi\left(1 - \dfrac{\delta^2}{6}\right)$	$\tan\chi\delta\cot\delta$ $= \tan\chi\left(1 - \dfrac{\delta^2}{3}\right)$	$x = \phi\sin\chi - l\phi\cos\chi - \dfrac{\phi^3}{6}\cos^2\chi\sin\chi + \dfrac{l\phi^3}{6}\cos^3\chi$ $\quad - \dfrac{\delta^2}{12}(6\phi\sin\chi - 2l\phi\cos\chi - 10\phi^3\cos^2\chi\sin\chi$ $\quad\quad\quad\quad + l\phi^3\cos^3\chi),$ $y = l + \dfrac{\sin 2\chi}{4}\phi^2 - \dfrac{\cos^2\chi}{2}l\phi^2 - \dfrac{\sin\chi\cos^3\chi}{24}\phi^4$ $\quad - \dfrac{\delta^2}{144}(24\sin 2\chi\phi^2 - 24\cos^2\chi l\phi^2$ $\quad\quad\quad\quad - 5\sin\chi\cos^3\chi\phi^4)$
Euler's projection for Russia	$\cos\chi\,\dfrac{\sin\delta}{\delta}$ $= \cos\chi\left(1 - \dfrac{\delta^2}{6}\right)$	$\dfrac{\delta}{2}\tan\chi\cot\dfrac{\delta}{2}$ $= \tan\chi\left(1 - \dfrac{\delta^2}{12}\right)$	$x = \phi\sin\chi - l\phi\cos\chi - \dfrac{\phi^3}{6}\cos^2\chi\sin\chi + \dfrac{l\phi^3}{6}\cos^3\chi$ $\quad - \dfrac{\delta^2}{72}(18\phi\sin\chi - 12l\phi\cos\chi - 7\cos^2\chi\sin\chi\phi^3$ $\quad\quad\quad\quad + 6l\phi^3\cos^3\chi),$ $y = l + \dfrac{\sin 2\chi}{4}\phi^2 - \dfrac{\cos^2\chi}{2}l\phi^2 - \dfrac{\sin\chi\cos^3\chi}{24}\phi^4$ $\quad - \dfrac{\delta^2}{96}(10\sin 2\chi\phi^2 - 16\cos^2\chi l\phi^2$ $\quad\quad\quad\quad - 3\cos^3\chi\sin\chi\phi^4)$
Errors of scale on extreme parallels equal and opposite to that on mean parallel	$\dfrac{2}{\delta}\tan\dfrac{\delta}{2}\cos\chi$ $= \cos\chi\left(1 + \dfrac{\delta^2}{12}\right)$	$\tan\chi\delta\cot\delta$ $= \tan\chi\left(1 - \dfrac{\delta^2}{3}\right)$	$x = \phi\sin\chi - l\phi\cos\chi - \dfrac{\phi^3}{6}\cos^2\chi\sin\chi + \dfrac{l\phi^3}{6}\cos^3\chi$ $\quad - \dfrac{\delta^2}{72}(18\phi\sin\chi - 6l\phi\cos\chi - \cos^2\chi\sin\chi\phi^3$ $\quad\quad\quad\quad - 3l\phi^3\cos^3\chi),$ $y = l + \dfrac{\sin 2\chi}{4}\phi^2 - \dfrac{\cos^2\chi}{2}l\phi^2 - \dfrac{\sin\chi\cos^3\chi}{24}\phi^3$ $\quad - \dfrac{\delta^2}{48}(3\sin 2\chi\phi^2 + 2\cos^2\chi l\phi^2)$
Murdoch's first projection	$\cos\chi$	$\tan\chi\,\dfrac{\sin\delta}{\delta}$ $= \tan\chi\left(1 - \dfrac{\delta^2}{6}\right)$	$x = \phi\sin\chi - l\phi\cos\chi - \dfrac{\phi^3}{6}\cos^2\chi\sin\chi + \dfrac{l\phi^3}{6}\cos^3\chi$ $\quad - \dfrac{\delta^2}{36}(6\phi\sin\chi - \phi^3\sin\chi\cos^2\chi),$ $y = l + \dfrac{\sin 2\chi}{4}\phi^2 - \dfrac{\cos^2\chi}{2}l\phi^2 - \dfrac{\sin\chi\cos^3\chi}{24}\phi^3$ $\quad - \dfrac{\delta^2}{144}(6\sin 2\chi\phi^2 - \cos^3\chi\sin\chi\phi^4)$
Murdoch's third projection	$\cos\chi\,\dfrac{\sin^2\delta}{\delta^2\cos\delta}$ $= \cos\chi\left(1 + \dfrac{\delta^2}{6}\right)$	$\tan\chi\delta\cot\delta$ $= \tan\chi\left(1 - \dfrac{\delta^2}{3}\right)$	$x = \phi\sin\chi - l\phi\cos\chi - \dfrac{\phi^3}{6}\cos^2\chi\sin\chi + \dfrac{l\phi^3}{6}\cos^3\chi$ $\quad - \dfrac{\delta^2}{36}(12\phi\sin\chi - 6l\phi\cos\chi + 2l\phi^3\cos^3\chi)$ $\quad\quad\quad\quad - \cos^2\chi\sin\chi\phi^3,$ $y = l + \dfrac{\sin 2\chi}{4}\phi^2 - \dfrac{\cos^2\chi}{2}l\phi^2 - \dfrac{\sin\chi\cos^3\chi}{24}\phi^3$ $\quad - \dfrac{\delta^2}{72}(12\cos^2\chi l\phi^2 + \cos^3\chi\sin\chi\phi^4)$
Minimum error projection	$\cos\chi\left(1 - \dfrac{\delta^2}{30}\right)$	$\tan\chi\left(1 - \dfrac{\delta^2}{5}\right)$	$x = \phi\sin\chi - l\phi\cos\chi - \dfrac{\phi^3}{6}\cos^2\chi\sin\chi + \dfrac{l\phi^3}{6}\cos^3\chi$ $\quad - \dfrac{\delta^2}{60}(14\phi\sin\chi - 2l\phi\cos\chi - 3\phi^3\cos^2\chi\sin\chi$ $\quad\quad\quad\quad + l\phi^3\cos^3\chi),$ $y = l + \dfrac{\sin 2\chi}{4}\phi^2 - \dfrac{\cos^2\chi}{2}l\phi^2 - \dfrac{\sin\chi\cos^3\chi}{24}\phi^3$ $\quad - \dfrac{\delta^2}{360}(24\sin 2\chi\phi^2 - 12\cos^2\chi l\phi^3$ $\quad\quad\quad\quad - 5\cos^3\chi\sin\chi\phi^4)$

may suppose $x = A\phi$, and it becomes necessary simply to find the value of A to produce the required results. If total errors on extreme parallels are made equal and opposite to that on the mid-parallel, we obtain

$$A = \cos^2\frac{\delta}{2} = 1 - \frac{\delta^2}{4} + \frac{\delta^4}{48}\cdots.$$

If the errors of scale on the extreme parallels are made equal and opposite to that on the mid-parallel,

$$A = \frac{\cos\delta}{\cos^2\dfrac{\delta}{2}} = 1 - \frac{\delta^2}{4} - \frac{\delta^4}{24}\cdots.$$

If we make the total area of the map true we find

$$A = \frac{\sin\delta}{\delta} = 1 - \frac{\delta^2}{6} + \frac{\delta^4}{120}\cdots.$$

Finally if we wish to find the total square error we must calculate

$$M = \int_\alpha^\beta (A\operatorname{cosec}\theta - 1)^2 d\theta$$
$$= 2\sin\delta - 4A\delta + A^2\log\frac{1 + \sin\delta}{1 - \sin\delta}.$$

Substituting the values of A from above, or merely putting $\chi = \dfrac{\pi}{2}$ in the results obtained for the conical projections, we have, for the plate carrée, when $A = 1$, $M = \dfrac{\delta^5}{10}$, for the first two above $\dfrac{7\delta^5}{120}$, and for the last $\dfrac{2\delta^5}{45}$, which agrees with the minimum value, given by

$$A = 1 - \frac{\delta^2}{6} - \frac{\delta^4}{72}\cdots.$$

Any of these values of A may be similarly applied to the Cassini projection, i.e. the transverse plate carrée, in order to reduce the error, but of course the conditions that produce the particular value will need corresponding changes, since the parallels and meridians to which these conditions apply are no longer geographical.

We cannot leave the consideration of the simple conic without a brief investigation of what is known as the polyconic projection, first used by the U.S. Coast Survey Office in 1855. Here the central meridian only is a straight line and divided truly. The parallels are all circular, but they are no longer concentric; each is drawn as though it were the standard parallel of a simple conic, and is divided truly, the remaining meridians being formed by joining the points of division.

In the simple conic with one standard parallel χ, the constant of the cone $n = \cos \chi$, and the semi-vertical angle of the cone which touches the sphere along the standard parallel is

$$\sin^{-1} n = \frac{\pi}{2} - \chi.$$

The radius of the standard parallel is thus $\tan \chi$. Hence in the simple polyconic each parallel of colatitude θ has a radius $\tan \theta$, and is developed from a cone of constant $\cos \theta$.

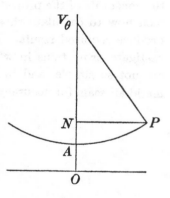

Now take the central meridian for axis of y, and the line through the centre of the map at right angles to it for that of x. Suppose P any point colatitude θ, longitude ϕ, and V_θ the vertex of the tangent cone belonging to the parallel AP. We then have

$$PV_\theta A = \phi \cos \theta, \quad PV_\theta = \tan \theta.$$

Thus the co-ordinates of P are

$$x = PN = \tan \theta \sin (\phi \cos \theta)$$
$$= \phi \sin \theta - \tfrac{1}{6} \phi^3 \sin \theta \cos^2 \theta + \tfrac{1}{120} \phi^5 \sin \theta \cos^4 \theta$$
$$= \phi \sin \chi - l\phi \cos \chi - \frac{\phi l^2}{2} \sin \chi + \frac{\phi l^3}{6} \cos \chi - \frac{\phi^3}{6} \sin \chi \cos^2 \chi \dots,$$

$$y = OA + AV_\theta - \tan \theta \cos (\phi \cos \theta)$$
$$= M + \tfrac{1}{2} \phi^2 \sin \theta \cos \theta - \tfrac{1}{24} \phi^4 \sin \theta \cos^3 \theta \dots,$$

where M is the meridian distance OA, or

$$y = l + \tfrac{1}{2}\phi^2 \sin\chi\cos\chi - \frac{\phi^2 l}{2}\cos 2\chi - \frac{\phi^4}{24}\sin\chi\cos^3\chi.$$

The advantage of this projection is that since the radius of each parallel depends only on its own colatitude and not on that of the centre, the map can be constructed in sheets having different centres, which when placed side by side fit approximately, as the curvature of the meridians is slight.

Leaving out this last projection, which though based on the simple conic is essentially not a simple conic, we may sum up the results of this chapter as follows. We have throughout been considering a projection in which, in the normal case, the meridian scale is correct; this makes the calculation of the radii of the parallels $(a + \theta)$ simple once the constants of the projection have been found, and we have seen how to calculate those constants in such a way as to produce required results. In the later chapters we shall investigate projections in which the expressions for the radii are not so simple, and in which we sacrifice correctness of meridian scale for accuracy in some other feature.

CHAPTER III

EQUAL AREA PROJECTIONS

As well as using the simple conic for a map of the world Ptolemy introduced a variation of it which is the earliest known projection that preserves the areas of small elements, and was the foundation of a much used one known afterwards as the projection of Bonne or the Carte de France. Instead of making all the meridians correct he took only one, the central meridian, passing through Meroë, divided that correctly and drew the parallels concentric circles as in the simple conic. Three of these, the equator and those of Thulé and Syene, he divided truly, and formed the remaining meridians by joining up the points of division of these parallels.

This method was used side by side with the simple conic for a number of maps constructed by the Arabs about A.D. 800 and later in 1511 by Bernard de Sylva, and from 1520 to 1566 by Pierre Benewitz, or Apianus, and Guillaume le Testu, who succeeded in dividing all the parallels correctly. In 1572 Bonne used it, and later de L'Isle, d'Anville and others, and in 1821 a commission under Laplace adopted it for the map of France in 69 sheets, considered in 1808, begun in 1818, and finished by L'Etât Major forty-five years later.

In this projection, as we have seen, the meridians are no longer straight lines; but it is easy to shew that the areas of small elements are correct. A small element of area on the map will be a parallelogram whose base is the element of the parallel, $\sin \theta \, d\phi$, since all the parallels are divided truly, and whose height is the element of the central meridian $d\theta$. Thus we have for the area $\sin \theta \, d\theta \, d\phi$, which is equal to the element of area on the earth—regarded as a sphere. Suppose V to be the common centre of the parallels, VA the central meridian and P any point, colatitude θ, longitude ϕ, measured from the central meridian. Then

$$VP = a + \theta,$$

since the parallels are described as in the simple conic, and arc

$$PA = \phi \sin\theta,$$

therefore

$$P\hat{V}A = \frac{\phi \sin\theta}{a+\theta}.$$

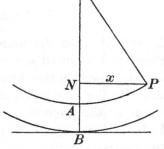

Hence the co-ordinates of P measured from axes through B, the intersection of the central meridian with the mean parallel χ, are

$$x = (a+\theta)\sin\left(\frac{\phi \sin\theta}{a+\theta}\right), \quad y = (a+\chi) - (a+\theta)\cos\left(\frac{\phi \sin\theta}{a+\theta}\right).$$

If we choose a, as is usual, as in the simple conic with one standard parallel χ, we have

$$x = \phi \sin\chi - l\phi \cos\chi - \frac{\phi^3}{6}\cos^2\chi \sin\chi - \frac{l^2\phi}{2}\sin\chi$$

$$+ \frac{l\phi^3}{6}\cos^3\chi + \frac{l^3\phi}{2}\cos\chi,$$

$$y = l + \frac{\phi^2}{2}\sin\chi \cos\chi - \frac{l\phi^2}{2}\cos^2\chi - \frac{\phi^4}{24}\sin\chi \cos^3\chi - \frac{l^2\phi^2}{2}\sin\chi \cos\chi,$$

where

$$l = \chi - \theta.$$

Notice that, to terms of the second degree, these are the same expansions as for the simple conic with one standard parallel, but also that, whereas in the simple conic l only appears to the first degree, in these expressions ascending powers of it occur.

A special case of this projection is that of Werner, who took $\chi = 0$ giving

$$x = \theta \sin\left(\frac{\phi \sin\theta}{\theta}\right) = \theta\phi - \frac{1}{6}\theta\phi^3 - \frac{1}{6}\phi\theta^3,$$

$$y = \theta \cos\left(\frac{\phi \sin\theta}{\theta}\right) = \theta - \frac{\theta\phi^2}{2} + \frac{\theta^3\phi^2}{6} + \frac{\theta\phi^4}{24} \ldots.$$

Another variation was used by Nicolas Sanson who, following the year 1650, constructed several maps on it, and also by Flamsteed for his star atlas begun in 1700 and finished in 1729. They took the equator for the standard parallel and obtained

$$x = \phi \sin \theta = \phi - \frac{l^2 \phi}{2} + \frac{l^4 \phi}{24} \dots ,$$

$$y = \frac{\pi}{2} - \theta = l.$$

This projection has for parallels a set of parallel straight lines at equal intervals from the equator, which is the axis of x; and it has been much used for maps of Africa, which is nearly bisected by the equator and is not of a very great extent of longitude.

The above three projections all satisfy the condition that the areas of small elements, and therefore total areas, are preserved everywhere on the map, and we will now proceed to investigate that condition in order to be able to tell by a simple calculation whether a pair of equations will give an equal area projection or not.

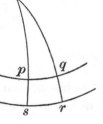

Suppose a small element $pqrs$ on the map bounded by two adjacent parallels θ, $\theta + \delta\theta$, and two adjacent meridians ϕ, $\phi + \delta\phi$. Then if p be the point x, y, q, r, s are the points

$$x + \frac{\partial x}{\partial \phi} \delta\phi, \quad y + \frac{\partial y}{\partial \phi} \delta\phi; \quad x + \frac{\partial x}{\partial \theta} \delta\theta + \frac{\partial x}{\partial \phi} \delta\phi,$$

$$y + \frac{\partial y}{\partial \theta} \delta\theta + \frac{\partial y}{\partial \phi} \delta\phi; \quad x + \frac{\partial x}{\partial \theta} \delta\theta, \quad y + \frac{\partial y}{\partial \theta} \delta\theta$$

respectively, and the area of the elementary quadrilateral $pqrs$ is

$$\pm \tfrac{1}{2} \left[\left(2y + \frac{\partial y}{\partial \phi} \delta\phi \right) \frac{\partial x}{\partial \phi} \delta\phi + \left(2y + \frac{\partial y}{\partial \theta} \delta\theta + 2 \frac{\partial y}{\partial \phi} \delta\phi \right) \frac{\partial x}{\partial \theta} \delta\theta \right.$$

$$\left. - \left(2y + 2 \frac{\partial y}{\partial \theta} \delta\theta + \frac{\partial y}{\partial \phi} \delta\phi \right) \frac{\partial x}{\partial \phi} \delta\phi - \left(2y + \frac{\partial y}{\partial \theta} \delta\theta \right) \frac{\partial x}{\partial \theta} \delta\theta \right]$$

$$= \pm \left(\frac{\partial x}{\partial \theta} \frac{\partial y}{\partial \phi} - \frac{\partial x}{\partial \phi} \frac{\partial y}{\partial \theta} \right) \delta\theta \, \delta\phi.$$

The area of the corresponding element of the spheroid is $\xi \delta\phi \delta\sigma$, and the equal area condition is

$$\frac{\partial x}{\partial \theta}\frac{\partial y}{\partial \phi} - \frac{\partial x}{\partial \phi}\frac{\partial y}{\partial \theta} = \pm \frac{(1 - \epsilon^2)\sin\theta}{(1 - \epsilon^2\cos^2\theta)^2}$$

$$= \pm (1 + 2e\cos 2\theta)\sin\theta,$$

or $\pm \sin\theta$, if we neglect the ellipticity.

The following solution of this condition is due to Germain, and gives the general form of the equations of an equal area projection, but it will often be found easier to choose the form of the equation beforehand and substitute in the condition to find the actual results.

Suppose we take x equal to some known function $F(\theta, \phi)$; then the general expression for y will be found by solving the equation

$$p\frac{\partial y}{\partial \phi} - q\frac{\partial y}{\partial \theta} = \pm (1 + 2e\cos 2\theta)\sin\theta,$$

where $$p = \frac{\partial F}{\partial \theta}, \quad q = \frac{\partial F}{\partial \phi}.$$

The integration of this equation necessitates the solution of the simultaneous equations

$$\frac{d\phi}{p} = -\frac{d\theta}{q} = \pm \frac{dy}{(1 + 2e\cos 2\theta)\sin\theta}.$$

These give $q\,d\phi + p\,d\theta = 0$, of which the integral is $F(\theta, \phi) = C$, where C is an arbitrary constant, and

$$dy = \pm \frac{(1 + 2e\cos 2\theta)\sin\theta}{q}\,d\theta.$$

If we suppose the equation $F(\theta, \phi) = C$ solved for ϕ and we substitute this value of ϕ in the expression for q, and obtain $q_{(\theta, c)}$ say, this quantity is no longer a function of ϕ, and theoretically it will be possible to find the integral of

$$\frac{(1 + 2e\cos 2\theta)\sin\theta}{q_{(\theta, c)}}\,d\theta,$$

and we can therefore write

$$y = C' \pm \int \frac{(1 + 2e \cos 2\theta) \sin \theta}{q_{(\theta, c)}} \, d\theta.$$

We then obtain the integral of the partial differential equations by fixing any arbitrary relation between C and C'; but since $F(\theta, \phi) = x$, the solution will be

$$x = F(\theta, \phi),$$

$$y = f(x) \pm \int \frac{(1 + 2e \cos 2\theta) \sin \theta}{\left(\dfrac{\partial x}{\partial \phi}\right)_{(\theta, c)}} \, d\theta,$$

where x is to be regarded as a constant in the expression $\dfrac{\partial x}{\partial \phi}$.

Take as an example

$$x = k(1 - e \sin^2\theta) \cdot \phi, \qquad \frac{\partial x}{\partial \phi} = k(1 - e \sin^2\theta),$$

and

$$\int \frac{(1 + 2e \cos 2\theta) \sin \theta}{\left(\dfrac{\partial x}{\partial \phi}\right)_{(\theta, c)}} \, d\theta = \int \frac{\sin \theta (1 + 2e \cos 2\theta)}{k(1 - e \sin^2\theta)} \, d\theta$$

$$= \frac{1}{k} \int \sin \theta (1 + 2e \cos^2\theta - e \sin^2\theta) \, d\theta$$

$$= -\frac{1}{k} \cos \theta (1 - e \sin^2\theta).$$

If, for simplicity, we suppress the $f(x)$, we have

$$y = \frac{1}{k} \cos \theta (1 - e \sin^2\theta),$$

giving a projection in which the parallels are a set of parallel straight lines and the meridians the hyperbolae

$$\frac{x^2}{k^2 \phi^2} - 2e k^2 y^2 = 1 - 2e,$$

i.e. a projection which, if we neglect the ellipticity, degenerates into a cylindrical one.

As in this last example, it sometimes happens that y, or x, is given in terms of θ only: we then have

$$q = \pm \frac{(1 + 2e \cos 2\theta) \sin \theta}{\dfrac{dy}{d\theta}},$$

and

$$x = \pm \phi \cdot \frac{(1 + 2e \cos 2\theta) \sin \theta}{\dfrac{dy}{d\theta}},$$

or for the sphere

$$\pm \phi \cdot \frac{\sin \theta}{\dfrac{dy}{d\theta}},$$

a result which we shall have occasion to use later on.

The conical projection of this class has been studied by many mathematicians, by Lambert in 1772, by Lorgna and Collignon later, but is connected chiefly with the name of Albers, who in 1805 used it for a general map of Europe published by Reichard at Nuremburg.

For a conical projection of any kind we must have

$$x = r \sin n\phi, \quad y = r \cos n\phi,$$

where r is a function of θ only. Substituting in the equal area condition for the spheroid without any approximation we have

$$nr \frac{dr}{d\theta} = -\frac{(1 - \epsilon^2) \sin \theta}{(1 - \epsilon^2 \cos^2 \theta)^2},$$

taking the negative sign for convenience. Therefore

$$r^2 = A - \frac{1}{n} \left\{ \frac{(1 - \epsilon^2) \cos \theta}{1 - \epsilon^2 \cos^2 \theta} + \frac{1 - \epsilon^2}{2\epsilon} \log \frac{1 + \epsilon \cos \theta}{1 - \epsilon \cos \theta} \right\},$$

where A and n are constants to be determined.

If we include only the first powers of e, we obtain

$$r^2 = a \left[c - \cos \theta \left(1 - 2e + \tfrac{4}{3} e \cos^2 \theta \right) \right],$$

where $a = \dfrac{2}{n}$ and $ac = A$.

Albers' projection has two standard parallels, θ_1 and θ_2 say, where $\theta_1 + \theta_2 = 2\chi$, $\theta_2 - \theta_1 = 2\delta$. Then

$$(1 + e\cos^2\theta_1)\sin\theta_1 = \frac{2}{a}\{a\,[c - \cos\theta_1\,(1 - 2e + \tfrac{4}{3}e\cos^2\theta_1)]\}^{\frac{1}{2}},$$

$$(1 + e\cos^2\theta_2)\sin\theta_2 = \frac{2}{a}\{a\,[c - \cos\theta_2\,(1 - 2e + \tfrac{4}{3}e\cos^2\theta_2)]\}^{\frac{1}{2}},$$

whence $\qquad n = \cos\chi\cos\delta\,\{1 + \tfrac{8}{3}e\delta^2\sin^2\chi\},$

and

$$c = \frac{\cos^2\chi + \cos^2\delta}{2\cos\chi\cos\delta} - \frac{e}{3\cos\chi\cos\delta}\{\cos^2\chi\,(3 - \cos^2\chi) - \delta^2\cos 2\chi\}.$$

For maps of the sphere we have

$$n = \cos\chi\cos\delta = \cos\chi\left(1 - \frac{\delta^2}{2}\cdots\right),$$

$$c - \cos\chi = \frac{\sin^2\chi}{2\cos\chi}\left(1 - \frac{\delta^2}{2}\cdots\right).$$

In the case of one standard parallel the expansions are

$$r = \tan\chi - l + \frac{l^3}{6} - \frac{5}{24}l^4\cot\chi\cdots,$$

$$x = \phi\sin\chi - l\phi\cos\chi - \frac{\phi^3}{6}\cos^2\chi\sin\chi + \frac{l\phi^3}{6}\cos^3\chi + \frac{l^3\phi}{6}\cos\chi\cdots,$$

$$y = l + \frac{\phi^2}{2}\sin\chi\cos\chi - \frac{l\phi^2}{2}\cos^2\chi - \frac{l^3}{6} + \frac{5l^4}{24}\cot\chi - \frac{\phi^4}{24}\sin\chi\cos^3\chi.$$

If instead of stipulating beforehand the two standard parallels we suppose that the scale errors on the extreme parallels are equal and opposite to those on the mid-parallel, we obtain

$$n = \cos\chi\left(1 + \frac{\delta^4}{16}(5 - 6\cot^2\chi)\right),$$

$$c - \cos\chi = \frac{\sin^2\chi}{2\cos\chi}\left(1 - \frac{\delta^2}{2}\cdots\right).$$

The determination of the total square error, by which we can compare projections of this kind, is laborious, but it can be done following Young's method of expansion in powers of δ. If we suppose

$$n = \cos\chi\,(1 + e\delta^2 + f\delta^4)$$

and

$$c - \cos\chi = \frac{\sin^2\chi}{2\cos\chi}\,(1 + g\delta^2 + h\delta^4),$$

and expand the integrals as far as terms in δ^5, we find that only e and g are required to be known and obtain the expression

$$M = \frac{\delta^5\sin\chi}{15}\{15\,(e+g)^2 + 10\,(e+g) + 3\},$$

giving for the above two cases the values

$$\tfrac{8}{15}\delta^5\sin\chi \quad \text{and} \quad \tfrac{7}{60}\delta^5\sin\chi,$$

indicating a considerable improvement if the standard parallels be situated more towards the middle of the map.

For the minimum error projection we must have

$$e + g = -\tfrac{1}{3},$$

and this condition will be found to be fulfilled if we make the scales on the extreme parallels equal and the total depth of the map true. These conditions give $e = \tfrac{1}{6}$ and $g = -\tfrac{1}{2}$, and for M we obtain the value $\tfrac{4}{45}\delta^5\sin\chi$, which is twice as great as in the absolute minimum error case.

Now let us pass on to the two limiting cases of the conical projection, viz. the cylindrical and the azimuthal. In the former we cannot at once put n equal to zero, but must go back to first principles and substitute $x = f(\phi)$, $y = F(\theta)$ in the general equal area condition. We then obtain

$$\frac{df}{d\phi}\cdot\frac{dF}{d\theta} = \sin\theta\,(1 + 2e\cos 2\theta).$$

Thus we must have

$$\frac{df}{d\phi} = k \quad \text{and} \quad f = k\phi + a,$$

$$\frac{dF}{d\theta} = \frac{1}{k}\sin\theta\,(1 + 2e\cos 2\theta)$$

giving $\quad F = b - \dfrac{1}{k}\left(\cos\theta - 2e\cos\theta\,(1 - \tfrac{2}{3}\cos^2\theta)\right).$

Therefore, taking the origin on the equator and at zero longitude we find

$$x = \frac{1}{k} \cos \theta \, (1 - 2e + \tfrac{4}{3} \, e \cos^2 \theta),$$

$$y = k\phi.$$

In the latter case we have χ and δ both zero, since there is one standard parallel, the pole. This gives $n = 1, c = 1 - \dfrac{2e}{3}$; for the sphere $r = 2 \sin \dfrac{\theta}{2}$, and the co-ordinates are

$$x = \theta\phi - \tfrac{1}{6} \theta\phi^3 - \tfrac{1}{24} \theta^3\phi,$$

$$y = \theta - \frac{\theta\phi^2}{2} - \frac{\theta^3}{24}.$$

This projection lends itself to a transverse application, and gives an interesting though, up to the present, little used projection of the whole sphere, due to Aitoff. We have

$$x^2 + y^2 = 4 \sin^2 \frac{\theta}{2},$$

$$\frac{x}{y} = \tan \phi,$$

and putting $\gamma = \dfrac{\pi}{2}$ in the last equation of p. 5 we find

$$\sin^2 \phi = \cot^2 \theta \tan^2 (\lambda - \zeta).$$

Substituting for θ and ϕ we have for the projected meridians the quartic curves

$$x^4 + x^2y^2\{1 + \sin^2(\lambda - \zeta)\} + y^4 \sin^2(\lambda - \zeta) - 4x^2$$
$$- 4y^2 \sin^2(\lambda - \zeta) + 4 \sin^2(\lambda - \zeta) = 0.$$

For the boundary of the hemisphere, i.e. $\lambda - \zeta = \dfrac{\pi}{2}$, this reduces to $(x^2 + y^2 - 2)^2 = 0$, a circle, and the other meridians are curves partly within and partly without this circle, as shewn in the figure, and all passing through the points $0, \pm \sqrt{2}$.

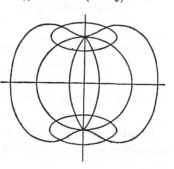

For the parallels we have from the second equation of p. 5

$$\cos^2 \mu = \sin^2 \theta \cos^2 \phi,$$

giving $x^2 y^2 + y^4 - 4y^2 + 4\cos^2 \mu = 0,$

a family of curves each consisting of two oval branches surrounding the poles, and cutting the axis of y at the points

$$\pm 2 \sin \left(\frac{\pi}{4} - \frac{\mu}{2} \right), \quad \pm 2 \cos \left(\frac{\pi}{4} - \frac{\mu}{2} \right).$$

For Aitoff's projection of the whole sphere only the part within the circle $x^2 + y^2 = 2$ is taken, and the graticule is projected orthogonally on to a plane through the equator $y = 0$, inclined to the original plane at an angle of 60°. This halves lengths parallel to the axis of y and the boundary of the hemisphere becomes an ellipse with major axis double the minor; if now we double the size of the whole map, and at the same time halve the scale of longitude we obtain a map of the whole world bounded by an ellipse of axes $2\sqrt{2}$ and $\sqrt{2}$ (area 4π), on which the parallels and meridians cut one another at angles never differing very greatly from right angles, and so give a fair representation of shape; and of course the projection is still equal area. It is easy to shew that the co-ordinates of any point are given by the expressions

$$x = \frac{2\sqrt{2}\sin \mu \sin \dfrac{\lambda - \zeta}{2}}{\left(1 + \sin \mu \cos \dfrac{\lambda - \zeta}{2} \right)^{\frac{1}{2}}} = 4 \sin \frac{D}{4} - \frac{l^2}{2} \sin \frac{D}{4} \left(1 - \tfrac{1}{2}\sec^2 \frac{D}{4} \right),$$

$$y = \frac{\sqrt{2}\cos \mu}{\left(1 + \sin \mu \cos \dfrac{\lambda - \zeta}{2} \right)^{\frac{1}{2}}} = l \sec \frac{D}{4} - \frac{l^3}{6} \sec \frac{D}{4} \left(1 - \tan^2 \frac{D}{4} \right),$$

where in this case l is the latitude $\dfrac{\pi}{2} - \mu$.

Similar in some respects to the last, though different in principle, is the projection of Mollweide (1805), which has been used for a map of the world, and of which many examples appear in most atlases. It has the great disadvantage that the distortion in high latitudes near the antipodes is very great, but apart from that it is very convenient, being neat in shape. The parallels are parallel straight lines and the meridians ellipses, and thus it is somewhat similar in appearance to the projection of Sanson, except that the parallels are not equidistant. For elliptic meridians we must suppose $x = a \cos t$, $y = b \sin t$, where a and b must be independent of the colatitude θ, and if the parallels are to be lines parallel to the axis of x, y must be a function of θ only and thus b is a constant and t independent of ϕ. Substituting in the equal area condition we obtain

$$b \frac{da}{d\phi} \cos^2 t \frac{dt}{d\theta} = - \sin \theta,$$

the negative sign being chosen for convenience. Since the left-hand side must be independent of ϕ,

$$a = \frac{k}{b} \phi \quad \text{and} \quad 2t + \sin 2t = \frac{4}{k} \cos \theta.$$

Thus $x = \frac{k\phi}{b} \cos t$, $y = b \sin t$, and the meridians are the ellipses

$$\frac{b^2 x^2}{k^2 \phi^2} + \frac{y^2}{b^2} = 1.$$

In Mollweide's projection the circular meridian is chosen for the boundary of the hemisphere; this gives $b^2 = \frac{k\pi}{2}$. Also since the area of the bounding ellipse, i.e. $k\pi^2$, must equal the surface area of the sphere, $k = \frac{4}{\pi}$ and $b = \sqrt{2}$. Thus we have

$$x = \frac{2\sqrt{2}\phi}{\pi} \cos t,$$

$$y = \sqrt{2} \sin t,$$

and t is given by

$$2t + \sin 2t = \pi \cos \theta,$$

or explicitly

$$t = \frac{1}{2}\frac{\pi\cos\theta}{2} + \frac{1}{24}\left(\frac{\pi\cos\theta}{2}\right)^3 + \frac{1}{120}\left(\frac{\pi\cos\theta}{2}\right)^5 + \frac{43}{20160}\left(\frac{\pi\cos\theta}{2}\right)^7.$$

We will call this expression $M(\theta)$. The boundary of the map is the ellipse

$$\frac{x^2}{8} + \frac{y^2}{2} = 1,$$

of which the major axis is double the minor. If the projection be applied transversely we obtain a beautiful map of the world also enclosed by a similar curve. The equator consists of the minor axis and the circumference of the bounding ellipse; the major axis is divided equally by the parallels; the boundary of the hemisphere is a circle radius $\sqrt{2}$, the meridians cut the equator at distances $\pm \sqrt{2}\sin M(\phi)$, where ϕ is the longitude, and the circular meridian cuts the parallels at points whose ordinates are $\pm \sqrt{2}\sin M(L)$ where L is the latitude.

About 1865 Collignon was studying equal area projections and he produced the equations of what we may conveniently call a trapezoidal projection, i.e. one in which the parallels are parallel and the meridians concurrent straight lines. This type of projection was much in evidence in earlier atlases, but according to Tissot, he was the first to give the equations of the equal area example. If the parallels are to be parallel to the axis of x we must have y independent of ϕ; and since the meridians are also straight lines we may suppose $y = b - ax$, where a and b are independent of θ.

Hence $\dfrac{dy}{d\theta} = -a\,\dfrac{dx}{d\theta}$, and from the equal area condition

$$x = +\frac{\phi\sin\theta}{a\dfrac{dx}{d\theta}} \quad \text{or} \quad x\,dx = +\frac{\phi}{a}\sin\theta\,d\theta,$$

of which the simplest solution is

$$x = 2c\phi \sin \frac{\theta}{2},$$

where $c = \dfrac{1}{\sqrt{a\phi}}$ and is independent of ϕ. Also

$$y = b - \frac{2}{c} \sin \frac{\theta}{2}.$$

If we take the origin on the equator, $b = \dfrac{\sqrt{2}}{c}$.

Collignon makes the meridian $\dfrac{\pi}{2}$ to be inclined at 45° to the parallels and therefore has $c = \sqrt{\dfrac{2}{\pi}}$, giving

$$x = 2\sqrt{\frac{2}{\pi}}\, \phi \sin \frac{\theta}{2},$$

$$y = \sqrt{\pi}\left(1 - \sqrt{2} \sin \frac{\theta}{2}\right).$$

If we took $c = \cos \dfrac{\chi}{2}$ the parallel of latitude χ would be a standard parallel and if

$$c = \cos \frac{\chi}{2}\left\{1 - \frac{\delta^2}{8}\left(2 \sec^2 \frac{\chi}{2} - 1\right)\right\},$$

the errors of scale on extreme parallels are equal and opposite.

Although this last is the only equal area trapezoidal projection, yet that of Prepetit Foucault has meridians which, for the part required, approximate to two straight lines intersecting on the equator. The equations are

$$x = \frac{2\phi}{\sqrt{\pi}} \sin \theta \cos^2 \left(\frac{\pi}{4} - \frac{\theta}{2}\right),$$

$$y = \sqrt{\pi} \tan \left(\frac{\pi}{4} - \frac{\theta}{2}\right).$$

In summing up the results of this chapter it is convenient to note two things; first that the equal area property pro-

duces the most elegant maps of the whole world, viz. Mollweide's normal and transverse and the transverse azimuthal
or Aitoff's; and secondly that, rather more than the others,
it gives curved meridians, and with simple equations. Those
of Bonne and Sanson appear with great frequency in most
atlases and in them the curvature of the meridians tends to
lessen the distortion in shape that is bound to follow ad-
herence to the equal area property. The conical projection of
this class is, as we have seen, complicated in form, and therefore is not so much in evidence and does not admit of so
many interesting variations as those of the others, and it
seems as though, for a conical projection, it were better to
follow Murdoch and make the total area of the map true,
allowing the small elements to obey some other law, perhaps
more useful and certainly less difficult in mathematical form.

CHAPTER IV

ORTHOMORPHIC PROJECTIONS

The two projections of Ptolemy which we have already mentioned, viz. the simple conic and the foundation of Bonne's, were temporarily superseded when, in 1569, Gerard Kauffman, or Mercator, produced his nautical chart on the projection giving the familiar map of the world that appears in all atlases. Here correctness of length and area give place to that of the shape of small elements, which is the principle of the orthomorphic or conformal projection, and in this particular case the projection is made first on a cylinder, as in that of the plate carrée, and the graticule consists of sets of orthogonal lines, of which the meridians are equidistant but the parallels so constructed as to make the two scales always equal. Since degrees of longitude increase as we near the equator, so the longitudinal scale, the meridians on the map being equidistant, increases as we approach the poles, and the degrees of latitude on the map are made to increase in the same ratio.

In 1599 Edward Wright published a table of latitudes giving numbers expressing the lengths of arcs of the meridians. Now the meridian scale of a cylindrical projection is $\dfrac{dx}{\cos L \, d\phi}$ and the parallel scale $\dfrac{dy}{dL}$; thus for this projection $\dfrac{dy}{dL} = \sec L$, since the meridians are equidistant. Hence Wright computed his table by the continued addition of the secants of $1''$, $2''$, $3''$, etc., as he was working before the days of the Integral Calculus, and it was not till about 1645 that Henry Bond noticed that this was a table of logarithmic tangents, while the actual demonstration of the definite integral,

$$\int_0^L \sec L \, dL = \log \tan \left(\frac{90 - L}{2} \right),$$

was first performed by James Gregory in 1668. Thus it was almost a century from the first construction of Mercator's map that the equation $y = \log \tan \frac{\theta}{2}$, giving the map ordinate in terms of the colatitude, was definitely proved.

Considerably more than another century passed before the master Gauss, who seems to have left no branch of mathematics unadorned by his researches, gave the general conditions for the conformal representation of one surface on another, that is, a representation in which scales and angles in every direction round a point are correct, though the scales vary with the point.

To find this condition suppose x, y, z are the co-ordinates of any point f on one surface, and X, Y, Z those of the corresponding point F on the other. Then the equations of the surfaces give x, y, z in terms of some parameters, t, u, say, and X, Y, Z in terms of some others, T, U. We have to find a relation between t, u and T, U to ensure the conformal similarity of the two surfaces, i.e. to ensure that any small element of the one surface is similar in shape to the corresponding element of the other.

Let g, h be two points

$$x + dx, y + dy, z + dz, \quad \text{and} \quad x + \delta x, y + \delta y, z + \delta z,$$

adjacent to f, and G, H the corresponding points $X + dX$, etc. adjacent to F. Then for the curvilinear triangle fgh to be similar to FGH we require

(1) $\dfrac{\text{arc } fg}{\text{arc } FG} = $ constant for all directions round f,

and   (2) $g\hat{f}h = G\hat{F}H$ for all directions of fg and fh.

Let   arc $fg = ds$,   arc $fh = \delta s$,   arc $FG = dS$,   arc $FH = \delta S$. Then the scale $m = \dfrac{\text{arc } FG}{\text{arc } fg} = \dfrac{dS}{ds}$, and this has to equal $\dfrac{\delta S}{\delta s}$.

From the equation to the first surface we have

$$dx = a\,dt + a'\,du,$$
$$dy = b\,dt + b'\,du,$$
$$dz = c\,dt + c'\,du,$$

where a, a', b, b', c, c' are definite functions of t, u. Now suppose we have found the required relations between the two surfaces, i.e. between t, u and T, U; they will enable us to express dX, dY, dZ in the form

$$dX = A\,dt + A'\,du,$$
$$dY = B\,dt + B'\,du,$$
$$dZ = C\,dt + C'\,du,$$

where A, A', B, B', C, C' are functions of t and u depending partly on the equation of the second surface, and partly on the relation between the two surfaces.

Hence we have

$$ds^2 = (a^2 + b^2 + c^2)\,dt^2 + 2\,(aa' + bb' + cc')\,dt\,.\,du$$
$$+ (a'^2 + b'^2 + c'^2)\,du^2,$$
$$dS^2 = (A^2 + B^2 + C^2)\,dt^2 + 2\,(AA' + BB' + CC')\,dt\,.\,du$$
$$+ (A'^2 + B'^2 + C'^2)\,du^2.$$

Since the scale m is to be the same for all directions it must be independent of the relation between t and u; and therefore we have

$$m^2 = \frac{dS^2}{ds^2} = \frac{A^2 + B^2 + C^2}{a^2 + b^2 + c^2} = \frac{AA' + BB' + CC'}{aa' + bb' + cc'} = \frac{A'^2 + B'^2 + C'^2}{a'^2 + b'^2 + c'^2}.$$

Also we have

$$ds\,\delta s \cos g\hat{f}h$$
$$= (a\,dt + a'\,du)\,(a\,\delta t + a'\,\delta u) + (b\,dt + b'\,du)\,(b\,\delta t + b'\,\delta u)$$
$$+ (c\,dt + c'\,du)\,(c\,\delta t + c'\,\delta u)$$
$$= (a^2 + b^2 + c^2)\,dt\,\delta t + (aa' + bb' + cc')\,(du\,\delta t + \delta u\,dt)$$
$$+ (a'^2 + b'^2 + c'^2)\,du\,\delta u,$$

and similarly

$$dS\,\delta S \cos G\hat{F}H$$
$$= (A^2 + B^2 + C^2)\,dt\delta t + (AA' + BB' + CC')$$
$$\times\,(du\,\delta t + \delta u\,dt) + (A'^2 + B'^2 + C'^2)\,du\,\delta u,$$

therefore $dS\,\delta S \cos G\hat{F}H = m^2 ds\,\delta s \cos g\hat{f}h.$

But $mds = dS$ and $m\delta s = \delta S$, therefore the condition that m is constant for all directions is sufficient to ensure that $g\hat{f}h = G\hat{F}H$. Now ds^2 is essentially positive and is a quadratic homogeneous function of dt and du; hence when resolved into factors it takes the form

$$ds^2 = n\,(dp + idq)\,(dp - idq),$$

where n is a finite real function of t and u, and dp and dq are real linear combinations of dt and du. It may be seen in all cases that dp and dq are perfect differentials, and thus we are at liberty to write them in that form.

Similarly we have

$$dS^2 = N\,(dP + idQ)\,(dP - idQ),$$

therefore $m^2 = \dfrac{N}{n}\dfrac{(dP + idQ)\,(dP - idQ)}{(dp + idq)\,(dp - idq)}.$

Now, from the expressions for ds^2 and dS^2 in terms of dt and du we see that the numerator and denominator of m^2 must, except for factors not involving infinitesimals, be the same. The conditions for this are that either

(1) $\dfrac{dP + idQ}{dp + idq}$ and $\dfrac{dP - idQ}{dp - idq}$,

or   (2) $\dfrac{dP + idQ}{dp - idq}$ and $\dfrac{dP - idQ}{dp + idq}$

are finite quantities simultaneously.

(1) If $\dfrac{dP + idQ}{dp + idq}$ is a finite quantity, $dP + idQ$ and $dp + idq$ must vanish simultaneously, and therefore $P + iQ$ and $p + iq$ must be constant together. Now P, Q, p, q are

all functions of t, u, hence $P + iQ$, $p + iq$, themselves variable quantities, cannot be constant together unless

$$P + iQ = f(p + iq),$$

where f is any function.

This implies two independent relations, since the real and imaginary parts of one side are respectively equal to those of the other. Hence we infer $P - iQ = f_1(p - iq)$, where $f_1 = f$ if i enters into f only in its occurrence in $p + iq$. Therefore if $P + iQ = f(p + iq)$, both the quantities

$$\frac{dP + idQ}{dp + idq} \quad \text{and} \quad \frac{dP - idQ}{dp - idq}$$

are finite simultaneously and the necessary conditions are fulfilled.

In exactly the same way condition (2) gives

$$P + iQ = f'(p - iq),$$

merely the same relation as above with the sign of q changed. Therefore to represent a surface orthomorphically on a plane we must first express the element of arc in the form given by $dS^2 = N^2(dP^2 + dQ^2)$ and then find the rectangular coordinates of any point on the plane from $x + iy = f(P + iQ)$, where f is to be chosen to suit any required conditions. The conical orthomorphic projection is found by making f an exponential function. For the spheroid the element of arc is given by

$$ds^2 = \xi^2 d\phi^2 + d\sigma^2, \quad \text{with the notation of Chap. I,}$$

$$= \xi^2 \left[\left(\frac{d\sigma}{\xi}\right)^2 + d\phi^2 \right] = \xi^2 (d\psi^2 + d\phi^2),$$

where $\qquad d\psi = \dfrac{d\sigma}{\xi} = \dfrac{(1 - \epsilon^2)\, d\theta}{(1 - \epsilon^2 \cos^2 \theta)\sin\theta},$

and therefore $\quad \psi = \log A \tan \dfrac{\theta}{2} \left(\dfrac{1 + \epsilon \cos\theta}{1 - \epsilon \cos\theta}\right)^{\epsilon/2}.$

Hence we suppose

$$ix + y = \exp n\,(\psi + i\phi)$$
$$= e^{n\psi}\,(\cos n\phi + i\sin n\phi),$$

and therefore

$$x = e^{n\psi}\sin n\phi,$$
$$y = e^{n\psi}\cos n\phi,$$

giving the radius of any parallel θ as

$$r = e^{n\psi} = A\tan^n\frac{\theta}{2}\left(\frac{1 + \epsilon\cos\theta}{1 - \epsilon\cos\theta}\right)^{n\epsilon/2}.$$

The constants A and n have now to be obtained in one of the various ways previously indicated. We will first suppose two standard parallels, which give what is known as Lambert's Second Projection; taking them as usual to be $\chi + \delta$ and $\chi - \delta$ we find

$$n = \frac{\log\dfrac{\sin(\chi + \delta)}{\sin(\chi - \delta)} - \dfrac{1}{2}\log\dfrac{1 - \epsilon^2\cos^2(\chi + \delta)}{1 - \epsilon^2\cos^2(\chi - \delta)}}{\log\dfrac{\tan\dfrac{\chi + \delta}{2}}{\tan\dfrac{\chi - \delta}{2}} + \dfrac{\epsilon}{2}\log\dfrac{[1 + \epsilon\cos(\chi + \delta)][1 - \epsilon\cos(\chi - \delta)]}{[1 - \epsilon\cos(\chi + \delta)][1 + \epsilon\cos(\chi - \delta)]}}$$

$$= \frac{\log\sin(\chi + \delta) - \log\sin(\chi - \delta) + e[\cos^2(\chi + \delta) - \cos^2(\chi - \delta)]}{\log\tan\dfrac{\chi + \delta}{2} - \log\tan\dfrac{\chi - \delta}{2} + 2e[\cos(\chi + \delta) - \cos(\chi - \delta)]},$$

neglecting higher powers of e, and

$$A = \frac{\sin(\chi - \delta)\cot^n\dfrac{\chi - \delta}{2}}{n\,[1 - \epsilon^2\cos^2(\chi - \delta)]^{\frac{1}{2}}}\left[\frac{1 - \epsilon\cos(\chi - \delta)}{1 + \epsilon\cos(\chi - \delta)}\right]^{n\epsilon/2}.$$

In the expansion of these expressions we follow the method of Young and use the general expression of p. 18. In that of n,

$$\log\sin(\chi + \delta) - \log\sin(\chi - \delta) = \int_{\chi - \delta}^{\chi + \delta}\cot\theta\,d\theta,$$

$$\log\tan\frac{\chi + \delta}{2} - \log\tan\frac{\chi - \delta}{2} = \int_{\chi - \delta}^{\chi + \delta}\operatorname{cosec}\theta\,d\theta,\ \text{etc.}$$

For n we obtain, as far as terms in δ^4 and $e\delta^2$,

$$n = \cos\chi\left[1 + \frac{\delta^2}{6}(1 + 8e\sin^2\chi) + \frac{\delta^4}{360}\left(\frac{16 + 7\sin^2\chi}{\sin^2\chi}\right)\right].$$

For the expansion of A and certain other expressions we here introduce the following set of functions. Suppose that for all integral values of r,

$$\frac{d^r}{d\chi^r}\left(\tan^n\frac{\chi}{2}\right) = K_r(\chi, n)\frac{\tan^n\frac{\chi}{2}}{\sin^r\chi},$$

then $K_r(\chi, n)$ is a rational integral function of $\cos\chi$ and n; for differentiating again we have

$$K_{r+1}(\chi, n)\frac{\tan^n\frac{\chi}{2}}{\sin^{r+1}\chi} = \left\{\frac{d}{d\chi}K_r(\chi, n)\right\}\frac{\tan^n\frac{\chi}{2}}{\sin^r\chi}$$

$$+ K_r(\chi, n)\left(\frac{n\tan^n\frac{\chi}{2}}{\sin^{r+1}\chi} - \frac{r\tan^n\frac{\chi}{2}\cos\chi}{\sin^{r+1}\chi}\right).$$

Therefore, writing K_r for $K_r(\chi, n)$,

$$K_{r+1} = \frac{dK_r}{d\chi}\sin\chi + nK_r - rK_r\cos\chi.$$

Therefore if K_r is a rational integral function of n and $\cos\chi$ so is K_{r+1}. But K_1 and K_2 are, therefore K_r is for all integral values of r.

The following values will be of use, and we give them here for reference:

$K_1(\chi, n) = n,$

$K_2(\chi, n) = n^2 - n\cos\chi,$

$K_3(\chi, n) = n^3 - 3n^2\cos\chi + n(1 + \cos^2\chi),$

$K_4(\chi, n) = n^4 - 6n^3\cos\chi + n^2(4 + 7\cos^2\chi).$

$$- n\cos\chi(5 + \cos^2\chi),$$

$K_5(\chi, n) = n^5 - 10n^4\cos\chi + 5n^3(2 + 5\cos^2\chi)$

$$- 5n^2\cos\chi(7 + 3\cos^2\chi) + n(5 + 18\cos^2\chi + \cos^4\chi),$$

$$K_1\{\chi, \cos\chi\,(1 + a\delta^2)\} \quad = \cos\chi + a\cos\chi\delta^2,$$

$$K_2\{\chi, \cos\chi\,(1 + a\delta^2)\} \quad = a\cos^2\chi\,.\,\delta^2,$$

$$K_3\{\chi, \cos\chi\,(1 + a\delta^2)\} \quad = \cos\chi\sin^2\chi - a\cos\chi\,(1 - 2\cos^2\chi)\,\delta^2,$$

$$K_4\,(\chi, \cos\chi) \qquad\qquad = -\cos^2\chi\sin^2\chi,$$

$$K_5\,(\chi, \cos\chi) \qquad\qquad = \cos\chi\sin^2\chi\,(5 - 2\cos^2\chi),$$

$$K_1\{\chi, 2\cos\chi\,(1 + a\delta^2)\} = 2\cos\chi + 2a\cos\chi\delta^2,$$

$$K_2\{\chi, 2\cos\chi\,(1 + a\delta^2)\} = 2\cos^2\chi\,(1 + 3a\delta^2),$$

$$K_3\{\chi, 2\cos\chi\,(1 + a\delta^2)\} = 2\cos\chi\sin^2\chi + 2a\cos\chi\,(1 + \cos^2\chi)\,\delta^2,$$

$$K_4\,(\chi, 2\cos\chi) \qquad\qquad = 6\cos^2\chi\sin^2\chi,$$

$$K_5\,(\chi, 2\cos\chi) \qquad\qquad = 2\cos\chi\sin^2\chi\,(5 - 7\cos^2\chi),$$

where powers of δ above the second have been neglected.

Also MacLaurin's theorem gives

$$\tan^n\frac{\chi - \delta}{2} = \tan^n\frac{\chi}{2}\left(1 - \frac{K_1\,(\chi, n)\,\delta}{\sin\chi} + \frac{K_2\,(\chi, n)}{\sin^2\chi}\frac{\delta^2}{2!}\cdots\right),$$

$$\cot^n\frac{\chi - \delta}{2} = \cot^n\frac{\chi}{2}\left(1 - \frac{K_1\,(\chi, -n)\,\delta}{\sin\chi} + \frac{K_2\,(\chi, -n)}{\sin^2\chi}\frac{\delta^2}{2!}\cdots\right).$$

Whence, differentiating logarithmically and expanding,

$$\frac{1}{[1 - \epsilon^2\cos^2(\chi - \delta)]^{\frac{1}{2}}} = \frac{1}{(1 - \epsilon^2\cos^2\chi)^{\frac{1}{2}}}\,(1 + e\sin 2\chi\delta - e\cos 2\chi\delta^2),$$

$$\left(\frac{1 - \epsilon\cos(\chi - \delta)}{1 + \epsilon\cos(\chi - \delta)}\right)^{n\epsilon/2} = \left(\frac{1 - \epsilon\cos\chi}{1 + \epsilon\cos\chi}\right)^{n\epsilon/2}(1 + e\sin 2\chi\delta - e\cos^2\chi\delta^2),$$

whence

$$A = \frac{\sin\chi\cot^n\dfrac{\chi}{2}}{n\,(1 - \epsilon^2\cos^2\chi)^{\frac{1}{2}}}\left(\frac{1 - \epsilon\cos\chi}{1 + \epsilon\cos\chi}\right)\left(1 - \frac{\delta^2}{2}\,(1 - 2e\sin^2\chi)\right).$$

To expand the radius r, we put $\theta = \chi - l$, and using

$$\left(\frac{1 + \epsilon\cos(\chi - e)}{1 - \epsilon\cos(\chi - e)}\right)^{n\epsilon/2} = \left(\frac{1 + \epsilon\cos\chi}{1 - \epsilon\cos\chi}\right)^{n\epsilon/2}\left[1 + e\sin 2\chi\left(1 + \frac{\delta^2}{6}\right)l\right.$$

$$\left. - e\cos^2\chi\left(1 + \frac{\delta^2}{6}\right)l^2\right],$$

obtain

$$r = A \tan^n \frac{\chi}{2} \left(\frac{1 + \epsilon \cos \chi}{1 - \epsilon \cos \chi} \right)^{n\epsilon/2} \Bigg[1 - l \cot \chi \left(1 - 2e \sin^2\chi \right.$$

$$\left. + \frac{\delta^2}{6}(1 + 6e \sin^2\chi) \right) - l^2 \cot^2\chi \left(3e \sin^2\chi - \frac{\delta^2}{12}(1 - 2e \sin^2\chi) \right)$$

$$- \frac{l^3}{6} \cot \chi \left(1 + 2e(1 - 4\cos^2\chi) + \frac{\delta^2}{6}\{1 - \cot^2\chi \right.$$

$$\left. + e(10 - 36\cos^2\chi)\} \right) \Bigg]$$

$$= \tan \chi \left(1 + e \cos^2\chi - \frac{\delta^2}{3}\{2 + e(1 + \cos^2\chi)\} \right)$$

$$\Bigg[1 - l \cot \chi \left(1 - 2e \sin^2\chi + \frac{\delta^2}{6}(1 + 6e \sin^2\chi) \right)$$

$$- l^2 \cot^2\chi \left(3e \sin^2\chi - \frac{\delta^2}{12}(1 - 2e \sin^2\chi) \right)$$

$$- \frac{l^3}{6} \cot \chi \left(1 + 2e(1 - 4\cos^2\chi) + \frac{\delta^2}{6}\{1 - \cot^2\chi \right.$$

$$\left. + e(10 - 36\cos^2\chi)\} \right) \Bigg].$$

From this we find for the co-ordinates, referred to axes through the intersection of the central meridian and the parallel χ,

$$x = \phi \sin \chi \left(1 + e \cos^2\chi - \frac{\delta^2}{2}\{1 + e(1 - 3\sin^2\chi)\} \right)$$

$$- \phi l \cos \chi \left(1 + e(1 - 3\sin^2\chi) - \frac{\delta^2}{3}\{1 + e(1 - 10\sin^2\chi)\} \right)$$

$$- \phi l^2 \cos \chi \cot \chi \left(3e \sin^2\chi - \frac{\delta^2}{12}\{1 + e(1 + 15\sin^2\chi)\} \right)$$

$$- \frac{\phi^3}{6} \sin \chi \cos^2\chi \left(1 + e \cos^2\chi - \frac{\delta^2}{6}\{1 + e(1 - 23\sin^2\chi)\} \right) \dots,$$

$$y = l \left(1 + e(1 - 3\sin^2\chi) - \frac{\delta^2}{2}\{1 + e(1 - 5\sin^2\chi)\} \right)$$

$$+ l^2 \cot \chi \left(3e \sin^2\chi - \frac{\delta^2}{12}\{1 + e(1 + 21\sin^2\chi)\} \right)$$

$$+ \frac{\phi^2}{2} \cos \chi \sin \chi \left(1 + e \cos^2\chi - \frac{\delta^2}{12}\{1 + e(1 - 8\sin^2\chi)\} \right)$$

$$- \frac{l\phi^2}{2} \cos^2\chi \left(1 + e(1 - 3\sin^2\chi) - \frac{\delta^2}{6}\{1 + e(1 - 25\sin^2\chi)\} \right)$$

$$+ \frac{l^3}{6} \left(1 - e(2 - 7\cos^2\chi) - \frac{\delta^2}{6}\{3 + \cot^2\chi \right.$$

$$\left. + e \cos^2\chi(5 + \cot^2\chi)\} \right) \dots.$$

These expressions reduce, in the case of the sphere, to

$$r = \tan \chi \left[1 - \tfrac{2}{3} \delta^2 - l \left(1 - \frac{\delta^2}{2} \right) + \frac{l^2 \delta^2}{12} \cot \chi \right.$$
$$\left. - \frac{l^3}{6} \left(1 - \frac{\delta^2}{6} (3 + \cot^2 \chi) \right) \right],$$

$$x = \phi \sin \chi \left(1 - \frac{\delta^2}{2} \right) - l\phi \cos \chi \left(1 - \frac{\delta^2}{3} \right) + \frac{\delta^2}{12} l^2 \phi \cos \chi \cot \chi$$
$$- \frac{\phi^3}{6} \sin \chi \cos^2 \chi \left(1 - \frac{\delta^2}{6} \right),$$

$$y = l \left(1 - \frac{\delta^2}{2} \right) - \frac{\delta^2 l^2}{12} \cot \chi + \frac{\phi^2}{2} \cos \chi \sin \chi \left(1 - \frac{\delta^2}{3} \right)$$
$$- \frac{l\phi^2}{2} \cos^2 \chi \left(1 + \frac{\delta^2}{6} \right) + \frac{l^3}{6} \left(1 - \frac{\delta^2}{6} (3 + \cot^2 \chi) \right).$$

And for the case of one standard parallel,

$$x = \phi \sin \chi - l\phi \cos \chi - \frac{\phi^3}{6} \sin \chi \cos^2 \chi \ldots,$$

$$y = l + \frac{\phi^2}{2} \sin \chi \cos \chi - \frac{l\phi^2}{2} \cos^2 \chi + \frac{l^3}{6} \ldots,$$

expressions identical with those previously obtained for the simple and equal area projections, as far as second degree terms.

Before we leave the case of the two standard parallels let us consider for a moment how, in the case of the sphere, the scale varies from point to point. Calling the scale m, we have, by comparing elements of length along the parallel θ,

$$m = \frac{nA \tan^n \dfrac{\theta}{2}}{\sin \theta},$$

whence
$$\frac{dm}{d\theta} = \frac{m}{\sin \theta} (n - \cos \theta),$$

and m is a minimum on the parallel $\cos^{-1} n$. To find the position of this parallel with reference to the two standard ones let us expand the numerator and denominator of n in terms of $c_1 = \cos(\chi - \delta)$, and $c_2 = \cos(\chi + \delta)$, both supposed positive. We have

$$n = \frac{\log \sin(\chi + \delta) - \log \sin(\chi - \delta)}{\log \tan \dfrac{\chi + \delta}{2} - \log \tan \dfrac{\chi - \delta}{2}}$$

$$= \frac{(c_1{}^2 + \tfrac{1}{2} c_1{}^4 + \tfrac{1}{3} c_1{}^6 \ldots) - (c_2{}^2 + \tfrac{1}{2} c_2{}^4 + \tfrac{1}{3} c_2{}^6 \ldots)}{2(c_1 + \tfrac{1}{3} c_1{}^3 + \tfrac{1}{5} c_1{}^5 \ldots) - 2(c_2 + \tfrac{1}{3} c_2{}^3 + \tfrac{1}{5} c_2{}^5 \ldots)}.$$

Here the denominator is positive since $c_1 > c_2$, and

$$n - c_2 = \frac{\displaystyle\sum_1^\infty \frac{1}{r}(c_1{}^{2r} - c_2{}^{2r}) - 2c_2 \sum_1^\infty \frac{1}{2r-1}(c_1{}^{2r-1} - c_2{}^{2r-1})}{2 \displaystyle\sum_1^\infty \frac{1}{2r-1}(c_1{}^{2r-1} - c_2{}^{2r-1})}$$

$$= \frac{\left\{ \begin{array}{l} \displaystyle\sum_1^\infty \frac{1}{r}(c_1{}^{2r-1} + c_1{}^{2r-2} c_2 + c_1{}^{2r-3} c_2{}^2 + \ldots c_2{}^{2r-1}) \\[2mm] \quad - \displaystyle\sum_1^\infty \frac{2c_2}{2r-1}(c_1{}^{2r-2} + c_1{}^{2r-3} c_2 + \ldots c_2{}^{2r-2}) \end{array} \right\}}{2 \displaystyle\sum_1^\infty \frac{1}{2r-1}(c_1{}^{2r-2} + c_1{}^{2r-3} c_2 + \ldots c_2{}^{2r-2})},$$

by dividing numerator and denominator by $c_1 - c_2$. The denominator is still positive, and the numerator is equal to

$$\sum_1^\infty \left\{ \frac{1}{r} c_1{}^{2r-1} + \left(\frac{1}{r} - \frac{2}{2r-1} \right)(c_1{}^{2r-2} c_2 + c_1{}^{2r-3} c_2{}^2 + \ldots c_2{}^{2r-1}) \right\}$$

$$= \sum_1^\infty \frac{1}{r(2r-1)} \big[(c_1{}^{2r-1} - c_1{}^{2r-2} c_2) + (c_1{}^{2r-1} - c_1{}^{2r-3} c_2{}^2) + \ldots$$
$$+ (c_1{}^{2r-1} - c_2{}^{2r-1}) \big],$$

where there are $2r - 1$ of the smaller brackets.

Now each of these small brackets is positive, therefore the whole sum is positive and

$$n - \cos(\chi + \delta) > 0, \text{ and } \cos^{-1} n < \chi + \delta.$$

Similarly $\cos^{-1} n > (\chi - \delta)$. Thus the parallel on which the scale is a minimum lies between the two standard parallels, and so the scale is too small between them and too great outside them. A similar investigation would shew that this fact is also true should either c_1 or c_2, or both of them, become negative.

In the case of one standard parallel the scale will obviously be correct on that parallel and too great elsewhere; and finally, the scale of areas, being equal to m^2, varies in a similar manner to m but to a much greater degree.

Now, before we pass on to the other methods of determining the constants, let us, as before, obtain a method of comparing these projections, by calculating the total square error in the general case. It will be found that in order to determine the first term of that expression only, it is not necessary to calculate the constant A further than the term in δ^2.

Suppose then, in any conical orthomorphic,

$$n = \cos \chi \, (1 + q\delta^2 + q' \, \delta^4),$$

$$A = \frac{\sin \chi \cot^n \dfrac{\chi}{2}}{n} (1 + p\delta^2 + p' \, \delta^4).$$

The total square error

$$M = \int_{\chi-\delta}^{\chi+\delta} 2 \left(\frac{nA \tan^n \dfrac{\theta}{2}}{\sin \theta} - 1 \right)^2 \sin \theta \, d\theta$$

$$= 4 \sin \chi \sin \delta + 2n^2 A^2 \int_{\chi-\delta}^{\chi+\delta} \frac{\tan^{2n} \dfrac{\theta}{2}}{\sin \theta} \, d\theta - 4An \int_{\chi-\delta}^{\chi+\delta} \tan^n \frac{\theta}{2} \, d\theta.$$

From the general expression for the expansion of an integral previously obtained we have, as far as terms in δ^5,

$$\int_{\chi-\delta}^{\chi+\delta} \frac{2n \tan^{2n} \dfrac{\theta}{2}}{\sin \theta} \, d\theta = \int_{\chi-\delta}^{\chi+\delta} \frac{K_1 (2n) \tan^{2n} \dfrac{\theta}{2}}{\sin \theta} \, d\theta$$

$$= \frac{2\delta \tan^{2n} \dfrac{\chi}{2}}{\sin \chi} \left[K_1 (2n) + \frac{K_3 (2n)}{\sin^2 \chi} \frac{\delta^2}{6} + \frac{K_5 (2n)}{\sin^4 \chi} \frac{\delta^4}{120} \cdots \right]$$

$$= \frac{4n\delta \tan^{2n} \dfrac{\chi}{2}}{\sin \chi} \left[1 + \frac{\delta^2}{6} + \frac{\delta^4}{120} (5 - 2 \cot^2\chi + 40q \cot^2\chi) \cdots \right],$$

and

$$\int_{\chi-\delta}^{\chi+\delta} \tan^n \frac{\theta}{2} \, d\theta = 2\delta \tan^n \frac{\chi}{2} \left[1 + \frac{\delta^4}{120} \cot^2 \chi \, (20q - 1) \ldots \right],$$

whence

$$M = 4\delta \sin \chi \left(1 - \frac{\delta^2}{6} + \frac{\delta^4}{120} \ldots \right)$$

$$+ 4\delta \sin \chi \left(1 + \frac{\delta^2}{6} + 2p\delta^2 + \frac{\delta^4}{120} \, (5 - 2 \cot^2 \chi \right.$$

$$\left. + 40q \cot^2 \chi + 240p' + 120p^2 + 40p) \right)$$

$$- 8\delta \sin \chi \left(1 + p\delta^2 + \frac{\delta^4}{120} \, (20q \cot^2 \chi - \cot^2 \chi + 120p') \right)$$

$$= \frac{\delta^5 \sin \chi}{15} \, (3 + 20p + 60p^2).$$

Now, therefore, when we have obtained the expression for A which makes an orthomorphic projection satisfy any conditions we like to choose, we can by substituting in this expression find out at once how the projection compares with others of the same class. For example, in the case of two standard parallels we have already found $p = -\frac{1}{2}$, and so we have for that projection

$$M = \tfrac{8}{15} \delta^5 \sin \chi.$$

Passing on to other methods of calculating the constants, let us suppose, as in Euler's equidistant projection, that the total errors on the two extreme parallels are equal to one another and opposite to that on the mean parallel χ.

This condition gives

$$nA \tan^n \frac{\chi - \delta}{2} - \sin (\chi - \delta) = \sin \chi - nA \tan^n \frac{\chi}{2}$$

$$= nA \tan^n \frac{\chi + \delta}{2} - \sin (\chi + \delta),$$

which reduce to the equation

$$K_1 - \cos\chi + \frac{\delta^2}{12}\left[\frac{2K_3}{\sin^2\chi} - \frac{3K_2\cos\chi}{\sin^2\chi} - 3K_1 + 2\cos\chi\right]$$

$$+ \frac{\delta^4}{240}\left[\frac{2K_5}{\sin^4\chi} - \frac{5K_4\cos\chi}{\sin^4\chi} - \frac{10K_3}{\sin^2\chi}\right.$$

$$\left. + \frac{10K_2\cos\chi}{\sin^2\chi} + 5K_1 - 2\cos\chi\right] = 0,$$

whence we find by successive approximation

$$n = \cos\chi\left(1 - \frac{\delta^2}{12} - \frac{\delta^4}{360}(7 + 29\cot^2\chi)\right),$$

giving

$$A = \frac{\sin\chi\cot^n\frac{\chi}{2}}{n}\left(1 - \frac{\delta^2}{4} - \frac{11\delta^4}{240}(1 + 2\cot^2\chi)\right),$$

and

$$M = \tfrac{7}{60}\delta^5\sin\chi,$$

which is exactly double the value obtained for Euler's map of Russia.

Again, if we suppose the errors of scale on extreme parallels equal and opposite to that on the mean, n has the same value as for two standard parallels, but

$$A = \frac{\sin\chi\cot^n\frac{\chi}{2}}{n}\left(1 - \frac{\delta^2}{4} + \frac{\delta^4}{48}(2 + 9\cot^2\chi)\right)$$

and

$$M = \tfrac{7}{60}\delta^5\sin\chi.$$

Thirdly, we might follow the method of Murdoch in his first projection, take $n = \cos\chi$ and make the total area true. This condition gives

$$nA^2\int_{\chi-\delta}^{\chi+\delta}\frac{2n\tan^{2n}\frac{\theta}{2}}{\sin\theta}\,d\theta = 4\sin\chi\sin\delta,$$

whence

$$A = \frac{\sin\chi\cot^n\frac{\chi}{2}}{n}\left(1 - \frac{\delta^2}{6} - \frac{\delta^4}{360}(1 - 3\cot^2\chi)\right),$$

and

$$M = \tfrac{4}{45}\delta^5\sin\chi.$$

Finally, suppose we take the expression

$$\tfrac{1}{15}(60p^2 + 20p + 3),$$

and write it in the form

$$4\{(p + \tfrac{1}{6})^2 + \tfrac{1}{45}\}.$$

We see at once that it is a minimum when $p = -\tfrac{1}{6}$ and that the value of M is then $\tfrac{4}{45}\delta^5 \sin\chi$ as in the last case. In a later chapter we shall discuss the absolute minimum error projection when it will be seen that this value of M is exactly double the minimum; but it is the least possible for an orthomorphic projection, and to obtain it we must choose the constant n and the other condition in such a way as to produce

$$A = \frac{\sin\chi \cot^n \dfrac{\chi}{2}}{n}\left(1 - \frac{\delta^2}{6}\right).$$

Now let us consider the two limiting cases of the conical projection, viz. the azimuthal and the cylindrical. In the former $n = 1$ and the parallels become complete circles of radii $2\tan\dfrac{\theta}{2}$, if we suppose the map to extend to the centre, since we must take the case of one standard parallel and that the pole. This is known as the Stereographic Projection, and it belongs to the class called Perspective Projections with which we deal in the next chapter. The meridian scale is $\sec^2\dfrac{\theta}{2}$ and increases rapidly as we leave the pole, so that the projection is of little use for maps of large areas but was at one time very popular in Germany. If we calculate the value of M for a region extending from the pole to the parallel β we obtain $\dfrac{\beta^6}{48}$, but by introducing a scale constant which we find to be $2\cot^2\dfrac{\beta}{2}\log\sec\dfrac{\beta}{2}$, so that the final expression for r is $4\cot^2\dfrac{\beta}{2}\log\sec\dfrac{\beta}{2}\tan\dfrac{\theta}{2}$, we can reduce the value of M to a minimum for an orthomorphic azimuthal projection, of

$\dfrac{\beta^6}{192}$; or if instead we use the much simpler factor $\cos\dfrac{\beta}{2}$, we diminish considerably the scale error on the extreme parallel, make the total area correct and obtain for M a value which agrees with this minimum as far as the first term.

The ordinary projection $r = 2\tan\dfrac{\theta}{2}$ lends itself readily to oblique or transverse application, and in all cases the parallels and meridians are circles whose centres, in the case of an oblique projection centred at a point γ, ζ, are respectively

$$0,\quad \dfrac{-2\sin\gamma}{\cos\mu+\sin\gamma}$$

and $\qquad -2\cot(\lambda-\zeta)\operatorname{cosec}\gamma,\quad 2\cot\gamma,$

and whose radii are

$$\dfrac{2\sin\mu}{\cos\mu+\cos\gamma}\quad\text{and}\quad 2\operatorname{cosec}(\lambda-\zeta)\operatorname{cosec}\gamma.$$

But we will return to these variations a little later.

Allied to the stereographic is the projection of Breusing, who suggested taking for the radius of a parallel the geometric mean between the values in this and the equal area case, giving

$$r = 2\left(\sin\dfrac{\theta}{2}\tan\dfrac{\theta}{2}\right)^{\frac{1}{2}}.$$

Young however has pointed out that a great improvement can be made by taking instead the harmonic mean, when r is found to be $4\tan\dfrac{\theta}{4}$. Both these may be modified by the introduction of a factor k; in the first case to make the total area true, we must take $k = \cot^2\dfrac{\beta}{2}$ and in the second $\cos^2\dfrac{\beta}{4}$, while in the latter, if k be taken equal to

$$\dfrac{\sin^2\dfrac{\beta}{4}}{\log\sec\dfrac{\beta}{2}-\tan^2\dfrac{\beta}{4}},$$

the total square error is reduced to a minimum.

To consider the cylindrical projections of this class it is best to return to the original condition $y + ix = f(\psi + i\phi)$, and make the function f merely a simple proportion. Thus we have

$$x = k\phi, \quad y = k\psi$$
$$= k \log \tan \frac{\theta}{2} \left(\frac{1 + \epsilon \cos \theta}{1 - \epsilon \cos \theta} \right)^{\frac{1}{2}}$$
$$= k \left[(1 - 2e) L + (1 + 2e) \frac{L^3}{6} \dots \right].$$

If there are two standard parallels θ_1 and θ_2,

$$k = \sin \theta_1 (1 + e \cos^2 \theta_1)$$
$$= \sin \theta_2 (1 + e \cos^2 \theta_2)$$

and $\qquad \theta_2 = \pi - \theta_1,$

as is otherwise obvious when we consider the fact that the cylinder on which the projection is first made cuts the spheroid along the standard parallels. In the case of one standard parallel that parallel must be the equator, and for the sphere we find again the expressions for Mercator's projection already mentioned. This projection is of the greatest value to navigators, for maps drawn on it have the property that any straight line on it represents a line of constant bearing on the earth. Since the projection is orthomorphic angles are everywhere preserved, and since the meridians are parallel the curve on the earth corresponding to any straight line on the map must cut every meridian at the same angle. If therefore a sailor wishes to sail from one point to another, keeping a constant bearing, he measures that bearing from a Mercator chart; he will not of course be sailing the shortest way, since the lines of constant bearing are not geodesics—but the question of what lines or curves represent the geodesics in different projections we must leave till a later chapter.

Mercator's projection may be modified to reduce the square error to a minimum, for which we must take

$$k = 1 - \frac{\delta^2}{6} + \frac{\delta^4}{72},$$

where 2δ is the depth of the map, giving

$$M = \frac{4\delta^5}{45}.$$

The transverse application is very interesting and we will consider it in some detail. Proposed by Gauss for a map of Egypt, it usually goes by the name of Gauss' Conformal Projection. Starting with the equations

$$x = \phi, \quad y = \log\tan\frac{\theta}{2}$$

for the sphere, we find that

$$\tan\frac{\theta}{2} = e^y, \quad \cos\theta = -\tanh y, \quad \sin\theta = \operatorname{sech} y,$$

and substituting in the equations for transverse projections we have for the parallels

$$\cos x = \cos\mu\cosh y,$$

and for the meridians

$$\sin x = -\tan(\lambda - \zeta)\sinh y.$$

The central meridian $\lambda - \zeta = \dfrac{\pi}{2}$ is the line $y = 0$ which cuts the parallel μ where $\cos x = \cos\mu$, and is thus divided truly. If we measure longitude from this line, we have $\zeta = -\dfrac{\pi}{2}$, and hence obtain

$$x = -\tan^{-1}(\tan\mu\cos\lambda),$$

$$y = \tanh^{-1}(\sin\mu\sin\lambda).$$

These co-ordinates may be expanded directly, or we may proceed as follows. The arbitrary relation between $x + iy$ and $\psi + i\lambda$ which determines the projection is found (cf. p. 66), when the axes of co-ordinates are suitably chosen, to be $x + iy = 2\tan^{-1}e^{\psi+i\lambda}$; and in spite of the fact that the projection is a transverse one, and that transverse projections are not formed for the spheroid as they are for the sphere, we may still suppose this relation to be satisfied and use it to determine the projection for the spheroid. Expanding the relation we have

$$x + iy = 2\left[e^{\psi+i\lambda} - \tfrac{1}{3}e^{3(\psi+i\lambda)} + \tfrac{1}{5}e^{5(\psi+i\lambda)} \ldots\right],$$

whence

$$x = 2\left[e^{\psi}\cos\lambda - \tfrac{1}{3}e^{3\psi}\cos 3\lambda + \tfrac{1}{5}e^{5\psi}\cos 5\lambda \ldots\right],$$

$$y = 2\left[e^{\psi}\sin\lambda - \tfrac{1}{3}e^{3\psi}\sin 3\lambda + \tfrac{1}{5}e^{5\psi}\sin 5\lambda \ldots\right].$$

If $\lambda = 0$, $\qquad x = 2\left[e^{\psi} - \tfrac{1}{3}e^{3\psi} + \tfrac{1}{5}e^{5\psi} \ldots\right],$

$$\frac{dx}{d\lambda} = \frac{d^3x}{d\lambda^3} = \frac{d^5x}{d\lambda^5} = \ldots = 0,$$

$$\frac{d^2x}{d\lambda^2} = -2\left[e^{\psi} - 3e^{3\psi} + 5e^{5\psi} \ldots\right],$$

$$\frac{d^4x}{d\lambda^4} = 2\left[e^{\psi} - 3^3e^{3\psi} + 5^3e^{5\psi} \ldots\right],$$

$$y = \frac{d^2y}{d\lambda^2} = \frac{d^4y}{d\lambda^4} = \ldots = 0, \quad \frac{dy}{d\lambda} = 2\left[e^{\psi} - e^{3\psi} + e^{5\psi} \ldots\right],$$

and so on.

To simplify the summation of these series let us suppose

$$f_n(z) = \left[z - 3^n z^3 + 5^n z^5 \ldots\right](1 + z^2)^{n+1}.$$

Then we find the law of formation

$$f_n(z) = z(1 + z^2)\frac{d}{dz}f_{n-1}(z) - 2nz^2 f_{n-1}(z),$$

and starting with $f_{-1}(z) = \tan^{-1}z$, we have

$$f_0(z) = z,$$

$$f_1(z) = z(1 - z^2),$$

$$f_2(z) = z(1 - 6z^2 + z^4), \text{ etc.}$$

We now have, when $\lambda = 0$ and n is even,

$$\frac{d^n x}{d\lambda^n} = \frac{(-)^{n/2} 2f_{n-1} (e^\psi)}{(1 + e^{2\psi})^n} = \frac{(-)^{n/2} 2f_{n-1} \left(\tan \frac{\mu}{2} (1 + 2e \cos \mu)\right)}{\left[1 + \tan^2 \frac{\mu}{2} (1 + 4e \cos \mu)\right]^n}$$

$$= (-)^{n/2} 2 \left[f_{n-1} \left(\tan \frac{\mu}{2}\right) \cos^{2n} \frac{\mu}{2} \right.$$

$$+ 2e \left\{ \tan \frac{\mu}{2} \left(1 + \tan^2 \frac{\mu}{2}\right) f'_{n-1} \left(\tan \frac{\mu}{2}\right) \right.$$

$$\left. - 2n \tan^2 \frac{\mu}{2} f_{n-1} \left(\tan \frac{\mu}{2}\right) \right\} \cos \mu \cos^{2n+2} \frac{\mu}{2} \right]$$

expanded to the first power of e, the ellipticity,

$$= (-)^{n/2} 2 \left[f_{n-1} + 2e \cos \mu \cos^2 \frac{\mu}{2} f_n \right] \cos^{2n} \frac{\mu}{2}.$$

Therefore

$$x = 2 \left[\left(f_{-1} + 2e \cos \mu \cos^2 \frac{\mu}{2} f_0 \right) \right.$$

$$- \frac{\lambda^2}{2} \cos^4 \frac{\mu}{2} \left(f_1 + 2e \cos \mu \cos^2 \frac{\mu}{2} f_2 \right)$$

$$\left. + \frac{\lambda^4}{24} \cos^8 \frac{\mu}{2} \left(f_3 + 2e \cos \mu \cos^2 \frac{\mu}{2} f_4 \right) \right]$$

$$= \mu + e \sin 2\mu - \frac{\lambda^2}{2} \sin \mu \cos \mu (1 + 2e \cos 2\mu)$$

$$+ \frac{\lambda^4}{24} \sin \mu \cos \mu [(1 - 6 \sin^2 \mu)$$

$$+ 2e (1 - 20 \sin^2 \mu + 24 \sin^4 \mu)].$$

Similarly

$$y = 2 \left[\left(f_0 + 2e \cos \mu \cos^2 \frac{\mu}{2} \right) \cos^2 \frac{\mu}{2} \lambda \right.$$

$$\left. - \left(f_2 + 2e \cos \mu \cos^2 \frac{\mu}{2} \right) \cos^6 \frac{\mu}{2} \cdot \frac{\lambda^3}{6} \cdots \right]$$

$$= \lambda \sin \mu (1 + 2e \cos^2 \mu)$$

$$- \frac{\lambda^3}{6} \sin \mu \{ \cos 2\mu + 2e \cos^2 \mu (1 - 6 \sin^2 \mu) \}$$

$$+ \frac{\lambda^5}{120} [\sin \mu (1 - 20 \sin^2 \mu + 24 \sin^4 \mu)$$

$$+ 2e \cos^2 \mu (1 - 60 \sin^2 \mu \cos 2\mu)].$$

In the case of the sphere we have

$$x = \mu - \frac{\lambda^2}{2} \sin \mu \cos \mu - \frac{\lambda^4}{4} \sin \mu \cos \mu \, (5 - 6 \cos^2 \mu),$$

$$y = \lambda \sin \mu - \frac{\lambda^3}{6} \sin \mu \cos 2\mu + \frac{\lambda^5}{120} \sin \mu \, (1 - 20 \sin^2 \mu + 24 \sin^4 \mu),$$

or in terms of the latitude, with origin on the equator,

$$x = l + \frac{\lambda^2}{2} \sin l \cos l + \frac{\lambda^4}{24} \sin l \cos l \, (5 - 6 \sin^2 l),$$

$$y = \lambda \cos l + \frac{\lambda^3}{6} \cos l \cos 2l + \frac{\lambda^5}{120} \cos l \, (1 - 20 \cos^2 l + 24 \cos^4 l),$$

and the value of x is seen to be the same as in the Cassini projection.

The above process is an example of a transverse projection applied to the spheroid, and it will be at once evident that such a method is possible only with an orthomorphic projection. The Gauss conformal has also been calculated for the spheroid in another way, viz. that of choosing certain properties that are fulfilled in the case of the sphere and making them also true in the case of the spheroid. Here we can stipulate the two conditions

(1) the projection is conformal, and

(2) the central meridian is divided truly.

Hence we must have (1) $x + iy = f\,(\psi + i\phi)$ and (2) the x axis is a meridian and distances along it are true. We must therefore select the x axis as the origin for longitude, and condition (2) becomes $x = \sigma$ (arc of meridian), when $y = 0$. Therefore we must have $x + iy = \sigma\,(\psi + i\phi)$, where $\sigma\,(\psi)$ is the expression for σ in terms of ψ. Expanding by Taylor's theorem and equating real and imaginary parts, we have

$$x = \sigma\,(\psi) - \frac{\phi^2}{2} \sigma_2\,(\psi) + \frac{\phi^4}{24} \sigma_4\,(\psi) \ldots,$$

$$y = \phi\sigma_1\,(\psi) - \frac{\phi^3}{6} \sigma_3\,(\psi) + \frac{\phi^5}{120} \sigma_5\,(\psi) \ldots,$$

where
$$\sigma_r\,(\psi) = \frac{d^r}{d\psi^r} \sigma.$$

Now $$d\psi = \frac{\rho}{\xi}\, d\theta \quad \text{and} \quad \rho = \frac{d\sigma}{d\theta}.$$

Therefore $$\frac{d\sigma}{d\psi} = \xi = \nu \sin\theta = \frac{1}{(1 - \epsilon^2 \cos^2\theta)^{\frac{1}{2}}},$$

and $$\frac{d\psi}{d\theta} = \frac{\rho}{\xi} = \frac{1 - \epsilon^2}{(1 - \epsilon^2 \cos^2\theta)\sin\theta}.$$

From these results we may obtain in succession the required differential coefficients and thus find, moving the origin to the equator,

$$x = M + \frac{\nu\phi^2}{2}\sin\theta\cos\theta + \frac{\nu\phi^4}{24}\sin^3\theta\cos\theta\,(5 - c^2 + 9\eta^2)\ldots,$$

$$y = \nu\phi\sin\theta + \frac{\nu\phi^3}{6}\sin^3\theta\,(1 - c^2 + \eta^2)$$

$$+ \frac{\nu\phi^5}{120}\,(5 - 18c^2 + c^4 + 14\eta^2 - 58c^2\eta^2)\ldots,$$

where M is the meridian distance of the point from the equator,

$$c = \cot\theta, \quad \eta = \frac{\epsilon^2 \sin^2\theta}{1 - \epsilon^2},$$

and we have neglected powers of η above the second.

In this projection, whichever method we adopt for the spheroid, the scales on the meridians and parallels will of course be equal. They are found from the expressions

$$\frac{1}{\rho}\left[\left(\frac{\partial x}{\partial\theta}\right)^2 + \left(\frac{\partial y}{\partial\theta}\right)^2\right]^{\frac{1}{2}}$$

and $$\frac{1}{\xi}\left[\left(\frac{\partial x}{\partial\phi}\right)^2 + \left(\frac{\partial y}{\partial\phi}\right)^2\right]^{\frac{1}{2}},$$

and as far as terms in ϕ^2 are, for the case of the sphere,

$$1 + \frac{\phi^2}{2}\sin^2\theta.$$

Now there is another projection which, though not orthomorphic, has meridian scale equal to this, and parallel scale very nearly equal to it, and actually equal to it on the central parallel. Suggested by Captain G. T. McCaw in a note read to the Royal Geographical Society on March 14th, 1921, it is a modification of what is known as the rectangular polyconic. The simple polyconic we have already considered, and the only difference in the rectangular case is that, instead of every parallel being divided truly, only the central one, usually the equator, though that is not necessary, is so divided, and through these points of division meridians are drawn so as to cut every parallel orthogonally.

Suppose P a point on a meridian θ, whose centre is V. Then

$$VP = \tan \theta, \text{ and } ON = \frac{\pi}{2} - \theta,$$

if O is the intersection of the central meridian and the equator. We may suppose

$$x = \tan \theta \sin \gamma,$$

$$y = \frac{\pi}{2} - \theta + \tan \theta - \tan \theta \cos \gamma,$$

where γ is a function of θ and ϕ to be determined so that the meridians and parallels are orthogonal. For this the tangent to the meridian at P must be VP, and therefore

$$\frac{\dfrac{\partial x}{\partial \theta}}{\dfrac{\partial y}{\partial \theta}} = -\tan \gamma,$$

whence $$\tan \frac{\gamma}{2} = f(\phi) \cos \theta,$$

where $f(\phi)$ must be determined so that the meridian meets the equator at the point ϕ, 0. This gives $f(\phi) = \frac{\phi}{2}$, and the expressions for the rectangular polyconic are thus

$$x = \frac{\phi \sin \theta}{1 + \dfrac{\phi^2}{4} \cos^2 \theta},$$

$$y = \frac{\pi}{2} - \theta + \tan \theta - \tan \theta \left(\frac{1 - \dfrac{\phi^2}{4} \cos^2 \theta}{1 + \dfrac{\phi^2}{4} \cos^2 \theta} \right)$$

$$= \frac{\pi}{2} - \theta + \frac{\phi^2 \sin \theta \cos \theta}{2 \left(1 + \dfrac{\phi^2}{4} \cos^2 \theta \right)}.$$

Captain McCaw worked on the idea that this might be made very nearly orthomorphic in the neighbourhood of a central parallel other than the equator. First he sought for a value of the function f which would make the projection orthomorphic. This expression, which to terms in ϕ^3 is

$$\frac{\phi}{2} \{1 + \tfrac{1}{12}\phi^2 (1 + \sin^2\theta)\},$$

will be seen to violate the conditions for the rectangular polyconic, since it contains θ, but if, instead, the value

$$\frac{\phi}{2} \{1 + \tfrac{1}{12}\phi^2 (1 + \sin^2\theta_0)\}$$

be taken, where θ_0 is the colatitude of the central parallel, then the scale errors on the meridians and parallels become respectively $\dfrac{\phi^2}{2} \sin^2\theta$, and $\dfrac{\phi^2}{4} (\sin^2\theta + \sin^2\theta_0)$, which are almost equal in the neighbourhood of the parallel θ_0, and on it are both actually equal to that in the Gauss conformal, while the common rectangular polyconic has a parallel scale error $-\dfrac{\phi^2}{4} \sin^2\theta$.

Another orthomorphic projection which has, for the sphere, the same scale error (as far as terms of the second degree) anywhere along the central parallel, is that of Lagrange, which, as we shall see in the last chapter, solves completely, for practical purposes, the problem of the orthomorphic projection of the spheroid. It is formed in a manner similar to that of the transverse conical orthomorphic, only instead of P being the actual point μ, D on the spheroid, it is a point on a sphere such that

$$\tan \frac{NP}{2} = e^{m(\psi + a)} \quad \text{and} \quad \angle C\hat{N}P = mD,$$

m and a being constants and C any fixed point on the equator. The co-ordinates are then given by

$$x = A \tan \frac{CP}{2} \sin \phi,$$

$$y = A \tan \frac{CP}{2} \cos \phi,$$

ϕ being the longitude measured from the meridian NC, with C as pole.

It follows at once that

$$x = \frac{A \sin CP \sin \phi}{1 + \cos CP} = \frac{A \sin NP \sin mD}{1 + \sin NP \cos mD}$$

$$= \frac{2A \sin mD}{\dfrac{t^m}{k} + 2 \cos mD + kc^m},$$

where

$$t^m = \frac{1}{c^m} = e^{m\psi},$$

and

$$k = e^{-ma}.$$

Similarly

$$y = \frac{A \left(\dfrac{t^m}{k} - kc^m \right)}{\dfrac{t^m}{k} + 2 \cos mD + kc^m}.$$

The constants A, m and k were determined by Lagrange so that the scale on the central meridian might be true at the centre of the map, and that its first and second differential coefficients might vanish everywhere, thus making the error on this meridian vary as the third power of the difference of latitude. We therefore find

$$kc_0{}^m = \frac{m + \cos \chi}{m - \cos \chi},$$

where χ is the colatitude of the central parallel and c_0 the value of c there,

$$m^2 = 1 + \sin^2\chi + \frac{2\epsilon^2}{1 - \epsilon^2}\sin^4\chi,$$
$$= 1 + \sin^2\chi + 2\epsilon^2\sin^4\chi,$$

though Lagrange neglected the ellipticity, and this value was only quite recently found by Young and McCaw almost simultaneously. We shall see later that this is not the best value to take for m, nor, so Young suggests, is it best to make the scale true at the centre and take

$$A = \frac{2m\xi_0}{m^2 - \cos^2\chi},$$

but to make it equal to one-third of and opposite to that on the extreme meridians. But with the above values the actual scale anywhere may easily be shewn to be

$$\frac{2Am}{\xi\left(\dfrac{t^m}{k} + 2\cos mD + kc^m\right)};$$

and on the central parallel we have to two terms

$$1 + \frac{\phi^2}{2}\sin^2\chi,$$

though we shall go into the question of the expansion of this quantity more fully in the last chapter.

The meridians and parallels in this projection will be found to be sets of coaxal circles with centres and radii

$$-A\cot mD, \quad 0; \quad A\operatorname{cosec} mD,$$

$$0, \quad \frac{A\left(\dfrac{t^m}{k} + kc^m\right)}{\dfrac{t^m}{k} - kc^m}; \quad A\left(\frac{kc^m}{\dfrac{t^m}{k} - kc^m}\right)^{\frac{1}{2}},$$

respectively; the poles are the points 0, $\pm A$, and the fundamental functional relation from which the co-ordinates can easily be derived is

$$y + ix = A \tanh m \, (\psi + \alpha + iD).$$

Now it is evident, both from the way in which the projection was derived, and from the form of the parallels and meridians, that the transverse stereographic for the sphere (p. 54) is a special case of Lagrange's projection, and it is therefore suggested that we should investigate the general transverse and oblique conical orthomorphic, and though we shall leave the expansion of it to the last chapter, yet we can find here the fundamental functional relation. We have in the normal case

$$y + ix = A \tan^n \frac{\theta}{2} \left(\cos n\phi + i \sin n\phi \right)$$

$$= A \left[\tan \frac{\theta}{2} \left(\cos \phi + i \sin \phi \right) \right]^n,$$

and supposing the pole of the map to be γ, ζ, we have

$$\tan \frac{\theta}{2} \left(\cos \phi + i \sin \phi \right) = \left(\frac{1 - \cos \theta}{1 + \cos \theta} \right)^{\frac{1}{2}}$$

$$\times \left(\frac{\cos \gamma \sin \mu \cos (\lambda - \zeta) - \sin \gamma \cos \mu + i \sin \mu \sin (\lambda - \zeta)}{\sin \theta} \right)$$

(where μ, λ are the geographical co-ordinates)

$$= \frac{\cos \gamma \operatorname{sech} \psi \cos D + \sin \gamma \tanh \psi + i \operatorname{sech} \psi \sin D}{1 - \cos \gamma \tanh \psi + \sin \gamma \operatorname{sech} \psi \cos D},$$

since $\quad \tan \dfrac{\mu}{2} = e^{\psi}, \quad \cos \mu = -\tanh \psi, \quad \sin \mu = \operatorname{sech} \psi,$

and where $D = \lambda - \zeta$. This fraction can be reduced by expressing the sine and cosine of γ in terms of $t = \tan \dfrac{\gamma}{2}$, and the functions of ψ and D as exponentials, and finally becomes

$$\frac{e^{\psi + iD} - t}{1 + t e^{\psi + iD}} = \tan \frac{f - \gamma}{2},$$

where $$\tan\frac{f}{2} = e^{\psi+iD}.$$

Therefore $$y + ix = A \tan^n \frac{f - \gamma}{2},$$

or in the cylindrical case

$$y + ix = A \log \tan \frac{f - \gamma}{2},$$

and for the Gauss conformal, putting $A = 1$, $\zeta = \frac{\pi}{2}$, and reversing the direction of y we find

$$e^{\psi+i\lambda} = \tan \frac{x + iy}{2}$$

as previously supposed.

In the oblique stereographic case $n = 1$, the co-ordinates will be found to be

$$x = \frac{2A \sin D \operatorname{cosec} \gamma}{\dfrac{t}{k} + 2 \cos D + kc},$$

$$y = A \frac{t + 2 \cot \gamma \cos D - c}{\dfrac{t}{k} + 2 \cos D + kc},$$

where $k = \cot \frac{\gamma}{2}$ and t and c are $\tan \frac{\mu}{2}$ and $\cot \frac{\mu}{2}$ respectively.

It has been suggested that this last is the best projection for a map of equal extent of latitude and longitude, and it is certainly true that angles are preserved everywhere, and that if $A = 2$ the scale is true at the centre in every direction, while if $A = 2 \cot^2 \frac{\beta}{2} \log \sec \frac{\beta}{2}$, as we have already seen, the total square error over a circular map of radius β is reduced to a minimum and the scale will be true on one particular circle.

Here we conclude, for the time being, the consideration of projections, all of which, with the exception of Breusing's and the rectangular polyconics, possess the orthomorphic property; and it is worthy of note that though this property may seem at first to be of the very utmost importance, yet it possesses two great disadvantages. The fact that the shapes of small elements are preserved does not necessarily mean that from looking at a map so constructed we can obtain a good idea of the shape of a large region. Areas, as in Mercator's map of the whole world, are often very greatly exaggerated, and of course distances are almost entirely incorrect; also the mathematical forms, though most elegant in theory, are as a rule complicated for practical purposes. Nevertheless the conical projection with two standard parallels (Lambert's) was used to a considerable extent during the Great War, and the stereographic was at one time very popular. This latter also belongs to the class of projections which are formed by actual geometrical projection from a point on to a plane or other figure, and these we will consider in the next chapter.

CHAPTER V

PERSPECTIVE PROJECTIONS

Although the earliest maps were constructed on the Carte Plate projection, and later ones on the two projections of Ptolemy, these were not the first known projections. Before the days of Thales and Anaximander there had appeared a geometrical projection of the sphere on to a tangent plane, either by lines radiating from the centre, called Gnomonic, or by parallel lines, called Orthographic, which has been attributed by some to Apollonius (B.C. 240) and by others to Hipparchus (B.C. 130), who also projected it from the point diametrically opposite to the point of contact of the plane, thus obtaining the stereographic; the use, however, of these projections in such early years was restricted to charts of the heavens, and it was not until 1514 that the last named was applied to geography by Werner of Nuremberg. These three projections, which are obviously azimuthal if the plane be a tangent at the north pole, and have the radii of their parallels proportional respectively to $\tan \theta$, $\sin \theta$ and $\tan \dfrac{\theta}{2}$, are all examples of perspective projections, and though originally this class was restricted to geometrical projections on to a tangent plane, that is not a necessary restriction, for in 1868 P. Braun suggested a perspective projection on to a cylinder, and in considering the general case we will suppose a cone placed with its axis lying along a diameter of the earth, and make a projection of points on the earth on to the surface of the cone from any point on that diameter. Then unwrapping the map from the cone into a plane we have a conical perspective projection.

Suppose V the vertex of the cone, here lying on the polar axis, and O the point from which the projection is made. Let P be any point on the spheroid, and let OP cut the cone in p. Then if the complement of the semivertical angle of the cone be β,

$$\frac{pm}{PM} = \frac{Om}{OM} = \frac{OV - pV \sin \beta}{OC + \eta}.$$

Let

$$OV = k, \quad OC = h, \quad Vp = r.$$

Then

$$r \cos \beta \, (h + \eta) = \xi \, (k - r \sin \beta),$$

whence

$$r = \frac{k \sin \theta}{h \, (1 - \epsilon^2 \cos^2 \theta)^{\frac{1}{2}} \cos \beta + \cos (\beta - \theta) - \epsilon^2 \cos \theta \cos \beta}$$

$$= \frac{k \sin \theta}{h \cos \beta + \cos (\beta - \theta)} \left(1 + \frac{e \cos \theta \cos \beta \, (h \cos \theta + 2)}{h \cos \beta + \cos (\beta - \theta)} \right).$$

Therefore since the constant of the cone is $\cos \beta$, we have, unwrapping into a plane map,

$$x = \frac{k \sin \theta \sin (\phi \cos \beta)}{h \cos \beta + \cos (\beta - \theta)} \left(1 + \frac{e \cos \theta \cos \beta \, (h \cos \theta + 2)}{h \cos \beta + \cos (\beta - \theta)} \right),$$

$$y = \frac{k \sin \theta \cos (\phi \cos \beta)}{h \cos \beta + \cos (\beta - \theta)} \left(1 + \frac{e \cos \theta \cos \beta \, (h \cos \theta + 2)}{h \cos \beta + \cos (\beta - \theta)} \right).$$

If the earth be regarded as a sphere the expression for the radius reduces to

$$\frac{k \sin \theta}{h \cos \beta + \cos (\beta - \theta)},$$

and has three constants to be determined. It will be found that the scales along parallels equidistant from β are equal— and if these scales are to be true $k = \sec \beta$—and thus it will be convenient, though of course not necessary, to take the parallel β for the central parallel of the map. Several interesting examples, three of which have already been mentioned, arise when $\beta = 0$, i.e. the central parallel is the pole and the projection is azimuthal. Until quite recently the plane of these azimuthal projections was always the tangent plane at the pole, and in that case $k = 1 + h$; this however is not at all necessary and by taking suitable values for k it is possible to reduce the error considerably.

The radius is now $\dfrac{k \sin \theta}{h + \cos \theta}$; if $h = 0$ we have the gnomonic projection obtained by drawing lines from the centre. On the tangent plane, $r = \tan \theta$, and the meridian scale is $\sec^2 \theta$, giving a very large error, but this projection has one very important property. Since every great circle on the sphere lies in a plane through the centre, every projected great circle is a straight line; at this point then we propose to digress for a moment to consider how in general to find the equation of a projected great circle.

Suppose A and B be two points on the sphere whose co-

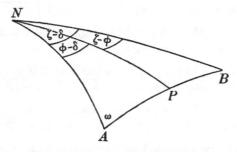

ordinates are γ, δ; η, ζ, respectively, and let P be a variable point θ, ϕ, on the great circle through A and B. Then

$$\cos AB \cos AP + \sin AB \sin AP$$

$$= \cos PB = \cos \theta \cos \eta + \sin \theta \sin \eta \sin (\zeta - \phi).$$

Whence, substituting for $\cos AB$, $\cos AP$ and $\sin AB$, $\sin AP$, using for the two latter

$$\cot \omega = \frac{\sin \gamma \cos \eta - \cos \gamma \sin \eta \cos (\zeta - \delta)}{\sin \eta \sin (\zeta - \delta)},$$

we eventually obtain for the relation between θ and ϕ any-where on the great circle

$$\tan \theta = \frac{\sin \gamma \sin \eta \sin (\zeta - \delta)}{\cos \gamma \sin \eta \sin (\zeta - \phi) + \sin \gamma \cos \eta \sin (\phi - \delta)}.$$

Thus to find the equation of a projected great circle, we must first express $\tan \theta$, $\sin \phi$, $\cos \phi$ in terms of the current co-ordinates x, y on the map, and substitute in this equation. But without even performing this operation we can find the direction of the projected great circle at either A or B, i.e. the true bearing of one from the other, in any conical projection.

If a be the projection of A, and α the angle between na and the tangent to the projected great circle,

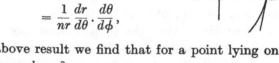

$$\cot \alpha = \frac{1}{nr} \frac{dr}{d\phi}$$

$$= \frac{1}{nr} \frac{dr}{d\theta} \cdot \frac{d\theta}{d\phi},$$

and from the above result we find that for a point lying on a great circle through γ, δ,

$$\frac{d\theta}{d\phi} = \sin \gamma \cot \omega,$$

when $\phi = \delta$, whence

$$\cot \alpha = \frac{1}{nr} \frac{dr}{d\theta} \sin \theta \cot \omega.$$

In the orthomorphic case of course, angles are preserved in the projection and $\alpha = \omega$, and in the case that suggested this investigation, the gnomonic, in which $r = k \tan \theta$, $n = 1$, we have

$$\cot \alpha = \frac{\cot \omega}{\cos \theta}, \quad \text{or} \quad \tan \omega = \sec \theta \tan \alpha.$$

This gives the bearing of B from A in terms of the angle α measured on the map, and in this case α is the angle nab, since the projected great circle is the straight line ab. It is

this last property of the gnomonic projection which no doubt
suggested the extension of it first produced by Reichard at
Weimar in 1803 in a representation of the whole sphere on
the faces of a circumscribed cube. If the cube be circumscribed normally, i.e. so that two faces touch at the north
and south poles, then on the other four equatorial faces the
meridians are the parallel lines $x = \tan(\lambda - \zeta)$, where ζ is
the longitude of the centre of the face, and the parallels the
hyperbolae
$$y^2 \tan^2 \mu - x^2 = 1,$$

these being the equations for the transverse gnomonic, and
the co-ordinate
$$y = \cot \mu \sec(\lambda - \zeta).$$

When two points are on the same face of the cube, we find
the great circle course between them simply by joining them
with a straight line, but when they are on adjacent or
opposite faces, it is not so simple. Suppose the cube laid out
flat and let aa', bb', cc' be the three lines of division of
adjacent faces; then the projection of any great circle PQ
must cut aa' as far above its middle point A, as it cuts cc'
below its middle point C. This can easily be seen by finding
its equations on the different faces, remembering that on the
north polar face the y co-ordinate is positive when measured
downwards. It will be found that in the case of P and Q
being on adjacent faces neither of which contains the pole
the length
$$BT = \frac{y_1 + y_2 + x_2 y_1 - x_1 y_2}{1 + x_1 x_2},$$

the co-ordinates being measured in each case from their re-
spective centres, i.e. $x_1 y_1$ from the centre of P's face, and
$x_2 y_2$ from that of Q's. This gives the distance from the mid
point of bb' to the point where the great circle cuts bb'; or
again we can construct the two parts of the line by the
following ingenious method due to Mr R. M. Milne of the
R.N.C., Dartmouth, and published in Hinks' *Map Projections*.

Draw PM, QN perpendicular to bb' and produce QN to Q' and M' making $M'Q' = Q'N = NQ$. Let MM' cut PQ' in X, and AX cut bb' in T. Join PT, TQ. Then if $HVXKW$

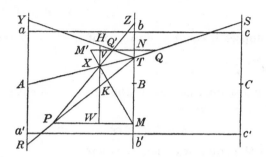

be parallel to bb', and if TQ' meet aa' in Y and XQ' meet bb' in Z, then

$$\frac{XK}{TZ} = \frac{PK}{PT} = \frac{PW}{PM} = \frac{Q'V}{Q'M'} = \frac{Q'V}{Q'N} = \frac{Q'H}{Q'T} = \frac{XH}{TZ}.$$

Therefore
$$XH = XK \quad \text{and} \quad AR = AY = SC.$$

Again, if P and Q are on opposite faces, first join P to Q', the point on the same face as P diametrically opposite to Q,

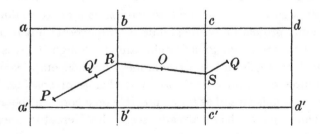

and continue the line round the faces of the cube with the same consideration as before, as shewn in the figure; for any great circle through Q' must pass through Q.

If instead of being circumscribed normally to the sphere,

the cube be placed so that the centre of one of its faces is at the point γ, ζ, then the parallels on this face are

$$x^2 \cos^2 \mu + y^2 (\cos^2 \mu - \sin^2 \gamma) - 2y \sin \gamma \cos \gamma$$
$$+ \cos^2 \mu - \cos^2 \gamma = 0,$$

and are ellipses if $\cos \mu > \sin \gamma$, i.e. if $\mu < \frac{\pi}{2} - \gamma$ and hyperbolae if $\mu > \frac{\pi}{2} - \gamma$.

The equator on this face is the line $y = - \cot \gamma$, and the meridians are the lines

$$x \cot (\lambda - \zeta) + y \cos \gamma = \sin \gamma,$$

which cut the equator at points where

$$x = \operatorname{cosec} \gamma \tan (\lambda - \zeta).$$

If the centres of three other faces also lie on the meridian ζ, which of course is not necessary but gives a convenient simplicity, then those of the two remaining are on the equator, and the equations of the parallels and meridians are exactly as in the normal case, but referred to axes inclined at an angle γ to the edges dividing them from the polar face. The drawing of a great circle may of course be proceeded with in exactly the same way as before.

It may seem that this great circle property must be of tremendous value in navigation, and thus tend to make the gnomonic projection correspondingly important; but it must be remembered that great circle sailing, though it gives the shortest route, is not the commonest method employed, as it requires constant calculation and change of bearing. The usual system is to sail as far as possible on a constant bearing, for which, as we have already seen, the Mercator chart is the most convenient.

The oblique application of the gnomonic projection used above suggests that the general plane perspective might also be applied obliquely, and it is easy to see from the simplest principles of geometrical projection, that on whatever plane

and from whatever point we project the parallels and meridians, they must project into curves of the first or second
degree. Applying our usual methods, we find, reversing the
direction of y, that the parallel μ is an ellipse whose centre
is the point

$$0,\ \frac{k\sin\gamma}{D^2}\,(h\cos\mu+\cos\gamma),$$

and whose semi-axes are

$$\frac{k\sin\mu}{D}\ \text{and}\ \frac{k\sin\mu\,(h\cos\gamma+\cos\mu)}{D^2},$$

where $D^2 = h^2 + 2h\cos\mu\cos\gamma + \cos^2\mu - \sin^2\gamma.$

Also the meridian λ is the ellipse centre

$$-\frac{k\sin\gamma\cot(\lambda-\zeta)}{E^2},\ -\frac{k\sin\gamma\cos\gamma}{E^2},$$

and semi-axes

$$\frac{hk\sin\gamma\,\text{cosec}\,(\lambda-\zeta)}{E^2},\ \frac{k\,\text{cosec}\,(\lambda-\zeta)}{E},$$

where $E^2 = (h^2-1)\cot^2(\lambda-\zeta) + h^2 - \cos^2\gamma,$

the equations of these axes being

$$x\cos\gamma - y\cot(\lambda-\zeta) = 0,$$

$$x\cot(\lambda-\zeta) + y\cos\gamma = -\frac{k\sin\gamma}{E^2}\{\cot^2(\lambda-\zeta)+\cos^2\gamma\}.$$

Now arises the important question of the best values to
take for h and k; once h is fixed the projection is obviously
determined in character, and variations in the value of k will
only serve to alter the errors. Besides the values 0, 1, ∞ of h
which give the gnomonic, stereographic and orthographic
projections respectively, the following values have been used:
$1 + \frac{1}{\sqrt{2}}$ by La Hire, 1·47 by Fiorini, and 1·367 by Sir Henry
James. These last two are different examples of the minimum
error projection of this class, first investigated by Col. Clarke,
who gives an account of it in his article on Map Projections

in the *Encyclopedia Britannica*. He suggested different values ranging between 1·65 and 1·35, which were found for different sized maps.

Let us suppose a perspective projection on a plane of a region extending from the pole to the parallel β. Then the total square error

$$M = \int_0^\beta \left\{ k^2 \left[\frac{(h \cos \theta + 1)^2}{(h + \cos \theta)^4} \sin \theta + \frac{\sin \theta}{(h + \cos \theta)^2} \right] \right.$$
$$\left. - 2k \left[\frac{h \cos \theta + 1}{(h + \cos \theta)^2} \sin \theta + \frac{\sin \theta}{h + \cos \theta} \right] + 2 \sin \theta \right\} d\theta.$$

Performing the integrations we have, using the notation of Col. Clarke

$$M = 4 \sin^2 \frac{\beta}{2} + 2kH + k^2 H',$$

where
$$H = \nu - (h + 1) \log (1 + \lambda),$$

$$H' = \frac{\lambda}{h + 1} (2 - \nu + \tfrac{1}{3}\nu^2),$$

and
$$\lambda = \frac{1 - \cos \beta}{h + \cos \beta}, \quad \nu = (h - 1) \lambda.$$

Differentiating M with respect to h and k for a minimum, we find

$$2 \frac{dH}{dh} + k \frac{dH'}{dh} = 0 \quad \text{and} \quad k = - \frac{H}{H'},$$

giving
$$2H' \frac{dH}{dh} - H \frac{dH'}{dh} = 0$$

as the equation for h, and for the minimum value of M

$$4 \sin^2 \frac{\beta}{2} - \frac{H^2}{H'}.$$

This last expression is the value of M for any projection made on to the minimum error plane, i.e. on to the plane given by $k = - \frac{H}{H'}$ whatever value be taken for h; and since H and H' contain h, the plane that gives the minimum error in any particular case varies with the value of h.

H and H' can most conveniently be expanded in powers of

$$c \left(= \frac{1}{h+1} \right) \text{ and } s \,(= 1 - \cos \beta),$$

for we then have

$$\lambda = \overset{\infty}{\underset{1}{\Sigma}}\, s^n c^n, \quad \nu = \overset{\infty}{\underset{1}{\Sigma}}\, (s^{n+1} - 2s^n)\, c^n, \quad \log (1 + \lambda) = \overset{\infty}{\underset{1}{\Sigma}}\, \frac{s^n c^n}{n},$$

giving

$$H = \overset{\infty}{\underset{1}{\Sigma}} \left(\frac{n}{n+1} s^{n+1} - 2s^n \right) c^n,$$

$$H' = \overset{\infty}{\underset{2}{\Sigma}} \left[\frac{n\,(n-1)}{6} s^{n+1} - \tfrac{1}{3}\, (n-1)\, (2n-1)\, s^n \right.$$
$$\left. + \tfrac{2}{3}\, (n^2 - 2n + 3)\, s^{n-1} \right] c^n.$$

Expanding H and H' as far as terms in s^3 and supposing

$$c = a + bs + \ldots,$$

we find

$$M = s^3 \,(\tfrac{4}{3}a^2 - a + \tfrac{5}{24}) = \frac{\beta^6}{8}\, (\tfrac{4}{3}a^2 - a + \tfrac{5}{24}).$$

Thus for the gnomonic ($c = 1$), $M = \dfrac{13\beta^6}{192}$, and for the stereographic ($c = \tfrac{1}{2}$), as we have already seen when regarding it as an orthomorphic projection, $M = \dfrac{\beta^6}{192}$.

If we differentiate with respect to a for the absolute minimum of M for perspective projections we find $M = \dfrac{\beta^6}{384}$, and $a = \tfrac{3}{8}$; and it appears that the first term in the expansion of M depends only on the constant term in the expansion of c, and that the value of h for a map of any size should be $\tfrac{5}{3}$. For greater accuracy however we must expand c to further terms. If we have any two expressions

$$H = a_1 x + a_2 x^2 + a_3 x^3 \ldots,$$
$$H' = \qquad\ \ b_2 x^2 + b_3 x^3 \ldots,$$

then the coefficient of x^n in the expansion of

$$2H' \frac{dH}{dx} - H \frac{dH'}{dx}$$

is
$$\sum_{r=1}^{n-1} (2n - 3r - 1) a_{n-r} b_{r+1}.$$

Substituting for the coefficients we find that in our case the complete expression is

$$\sum_{n=1}^{\infty} c^n \left[\{ \tfrac{1}{6}(n+1)(n+2)(n+4)U_n - \tfrac{1}{72}n(n+1)(n-1)(n-2) \right.$$
$$- \tfrac{1}{12}(n-1)(n+2)(5n+8)\} s^{n+3}$$
$$+ \left\{ \frac{(n-1)}{12}(n^3 + 19n^2 + 60n + 48) \right.$$
$$- \tfrac{1}{3}(2n^3 + 13n^2 + 23n + 12)U_n \} s^{n+2}$$
$$+ \{ \tfrac{2}{3}(n+4)(n^2 + 2n + 3)U_n$$
$$- \tfrac{1}{6}(n-1)(n^3 + 9n^2 + 20n + 48)\} s^{n+1}$$
$$\left. + \tfrac{1}{9}(n-1)(n-2)(n-3)(n+4)s^n \right],$$

where
$$U_n = \sum_2^n \frac{1}{r}.$$

Expanding this as far as $n = 8$ and dividing by $s^4 c^3$, we find for the equation in h

$$-2 + s - \frac{s^2}{18} + c\left(\tfrac{16}{3} - 12s + \tfrac{46}{9}s^2 - \tfrac{1}{3}s^3\right)$$
$$+ c^2\left(24s - \tfrac{197}{5}s^2 + \tfrac{157}{10}s^3 - \tfrac{23}{20}s^4\right)$$
$$+ c^3\left(\tfrac{200}{3}s^2 - 97s^3 + \tfrac{75}{2}s^4 - \tfrac{73}{3}s^5\right)$$
$$+ c^4\left(\tfrac{440}{3}s^3 - \tfrac{7037}{35}s^4 \ldots\right) + c^5\left(280s^4 \ldots\right) = 0.$$

And rearranging in powers of s as far as s^4 we find

$$\tfrac{16}{3}c - 2 + s(1 - 12c + 24c^2) + s^2\left(-\tfrac{1}{18} + \tfrac{46}{9}c - \tfrac{197}{5}c^2 + \tfrac{200}{3}c^3\right)$$
$$+ s^3\left(-\tfrac{1}{3}c + \tfrac{157}{10}c^2 - 97c^3 + \tfrac{440}{3}c^4\right)$$
$$+ s^4\left(-\tfrac{23}{20}c^2 + \tfrac{75}{2}c^3 - \tfrac{7037}{35}c^4 + 280c^5\right) = 0.$$

Solving this for c by successive approximation we have

$$c = \frac{3}{8} + \frac{3s}{128} + \frac{67s^2}{15360} + \frac{193s^3}{163840} + \frac{153469s^4}{412876800} + \cdots,$$

from which we obtain

$$h = \frac{5}{3}\left(1 - \frac{s}{10} - \frac{89s^2}{7200} - \frac{89s^3}{28800} - \frac{542809}{154828800}\right),$$

or in terms of β,

$$h = \frac{5}{3}\left(1 - \frac{\beta^2}{20} + \frac{31\beta^4}{28800} \cdots\right),$$

$$k = \frac{8}{3}\left(1 - \frac{3\beta^2}{32} - \frac{29\beta^4}{46080} \cdots\right).$$

For Fiorini's projection of the hemisphere we obtain, by putting $\beta = \frac{\pi}{2}$, $h = 1 \cdot 471$ and $k = 2 \cdot 04$. Sir Henry James took $\beta = 113\frac{1}{2}°$ giving $h = 1 \cdot 367$, $k = 1 \cdot 66$. Clarke obtained the value $1 \cdot 625$ for h for a map of South Africa or South America for which $\beta = 40°$, using a method of trial and error, while the expressions above give $h = 1 \cdot 626$ and $k = 2 \cdot 54$.

Now let us turn for a moment to the case of the general conical perspective projection for a zone. We should find M to be of the form

$$4 \sin \chi \sin \gamma + 2kH + k^2 H',$$

and could proceed exactly as above, for the integration of the terms of M is possible but extremely laborious. If however we attempt to expand the integrals in terms of the mean value χ and the semi-difference δ between the bounding parallels, putting

$$\cos \beta = \cos \chi (1 + c\delta^2) \quad \text{and} \quad h = a + b\delta^2,$$

we find that there appears in the second differential co-efficient of

$$\frac{(h \cos \theta + 1)^2 \sin \theta}{\{h \cos \beta + \cos (\theta - \beta)\}^4},$$

a term in $a^2 \sin^2 \theta$, and therefore, to get rid of the coefficient of δ^3 it becomes necessary to suppose $a = 0$; in this case we find $M = \dfrac{\delta^5 \sin \chi}{4}$, which, though a better result than that obtained in the case of the plane gnomonic or even stereographic does not compare favourably with that of Clarke's projection.

If the projection be actually from the centre, i.e. $b = 0$ as well as a, giving a conical gnomonic, the expression for the radius becomes $k \sin \theta \sec (\beta - \theta)$, and if the scale on extreme parallels are equal, i.e. $\beta = \chi$, then we find

$$k = \sec \chi \left(1 - \frac{\delta^2}{4} - \frac{13\delta^4}{240}\right).$$

Thus it does not appear that much advantage, if indeed any, can be obtained by projecting on to a cone rather than on to a plane: and the same remark is also true of the projection on the cylinder, which we have already mentioned as devised by Braun. However it is worth a passing glance.

Supposing a normal projection we should find quite readily

$$x = k\phi,$$

$$y = \frac{k (h + \cos \theta)}{\sin \theta} - h,$$

where k is the radius of the cylinder, and h as in the other perspective projections. For standard parallels χ and $\pi - \chi$, $k = \sin \chi$, and we may determine b so as to make the meridian scale correct on one of these parallels, giving

$$h = \mp \tan \left(\frac{\pi}{4} - \frac{\chi}{2}\right).$$

But the method of minimum error seems to apply even less well to this projection than to the general conical case, and we need not consider it further. In fact all these perspective projections, although greatly praised by Col. Clarke in his article as being the only purely geometrical ones, have been very little used even after the improvements obtained by applying the principle of minimum error, and in the next chapter we shall consider this principle further, and investigate what expressions, in the case of conical and azimuthal projections, give the absolute minimum error. Up to the present we have been using the principle merely to calculate the best constants, now we are to use it to find the best functions.

CHAPTER VI

MISCELLANEOUS PROJECTIONS

I. The Minimum Error Conical Projection

The principle of minimum error was originally devised in 1861 by Sir George Airy, who found an expression for the radius of a parallel in an azimuthal projection which would make the total square error, calculated for a map extending to the pole, a minimum; he omitted, however, to apply the principle to the calculation of the constant multiplier that occurs, taking it instead as unity, and so failed to find the expression giving the absolute minimum error for such an expression. The work of Mr A. E. Young on this branch of our subject has gone much further than that of its originator, for not only has he corrected Airy's expression, but he has also extended the method to the general conical case for the maps of a zone and of a portion of the earth's surface extending from the pole to any parallel; and the following investigation of the projection is taken almost bodily from that pamphlet of his to which many references have already been made.

Let the reader again be reminded that the minimum error principle has up to the present been rigidly applied only to the sphere and even then its application has not produced, except in one case, a projection that can be put to practical use. We shall find the expression for the radius of the general parallel but in a form that does not admit of expansion in terms of any known functions, and which does not serve any direct purpose. But it is valuable in that from it we are able to find the absolute minimum value of the total square error possible for a conical projection, or at any rate the beginning of its expansion; and though we cannot use directly the expression for the radius, yet, knowing this minimum value, we can be fairly sure that when we find another such expression which gives a value for M approximating to the known minimum, then that expression is close to the minimum error one. This we shall find to be true of the two projections of Murdoch and of the minimum error projection with true meridian scale (p. 20).

As we have seen in Chapter I, for the map of a zone extending between the parallels α and β

$$M = \int_\alpha^\beta \left[\left(\frac{nr}{\sin \theta} - 1 \right)^2 + \left(\frac{dr}{d\theta} - 1 \right)^2 \right] \sin \theta \, d\theta,$$

r being the radius of the parallel θ in a projection on a cone of constant n, i.e.

$$M = \int_\alpha^\beta V d\theta, \text{ say.}$$

Then, by the Calculus of Variations, we must have, for M to be a minimum,

$$\frac{\partial V}{\partial r} - \frac{d}{d\theta} \left(\frac{\partial V}{\partial p} \right) = 0,$$

where $\qquad p = \dfrac{dr}{d\theta}, \text{ and } \dfrac{\partial V}{\partial p} = 0$

when $\theta = \alpha$ or β.

The first condition gives

$$n \left(\frac{nr}{\sin \theta} - 1 \right) - \frac{d}{d\theta} \left\{ \left(\frac{dr}{d\theta} - 1 \right) \sin \theta \right\} = 0,$$

which reduces to the equation

$$\sin^2 \theta \frac{d^2 r}{d\theta^2} + \sin \theta \cos \theta \frac{dr}{d\theta} - n^2 r = \sin \theta \cos \theta - n \sin \theta.$$

To solve this equation, first substitute for θ in terms of $t \left(= \tan \dfrac{\theta}{2} \right)$ and then put $r = ut^n$. This produces

$$t^{2n+1} \frac{d^2 u}{dt^2} + (2n + 1) t^{2n} \frac{du}{dt} = \frac{2t^n (1 - t^2)}{(1 + t^2)^2} - \frac{2nt^n}{1 + t^2},$$

and integrating, we have

$$u = A + Bt^{-2n} + t^{-2n} \int \frac{2t^n}{1 + t^2}\, dt,$$

whence

$$r = A \tan^n \frac{\theta}{2} + B \cot^n \frac{\theta}{2} + \cot^n \frac{\theta}{2} \int \tan^n \frac{\theta}{2}\, d\theta,$$

where A and B are constants.

The only case in which $\int \tan^n \frac{\theta}{2}\, d\theta$ can be evaluated absolutely in terms of known functions is when $n = 1$ (the azimuthal), for though we have

$$\tan^n \frac{\theta}{2} = \frac{d}{d\theta} \tan^n \frac{\theta}{2} \left[\theta - \frac{K_1\,(n_1\theta)}{\sin\theta} \frac{\theta^2}{2!} + \frac{K_2\,(n_1\theta)}{\sin^2\theta} \frac{\theta^3}{3!} \cdots \right],$$

giving

$$\int \tan^n \frac{\theta}{2}\, d\theta = \theta \tan^n \frac{\theta}{2} \left[1 - \frac{K_1}{2!} \frac{\theta}{\sin\theta} + \frac{K_2}{3!} \left(\frac{\theta}{\sin\theta} \right)^2 \cdots \right],$$

this expansion, as can readily be seen, is of no value. When $n = 1$ we find

$$r = A \tan \frac{\theta}{2} + B \cot \frac{\theta}{2} + 2 \cot \frac{\theta}{2} \log \sec \frac{\theta}{2},$$

and to determine A and B we must use the second condition, which gives

$$\left(\frac{dr}{d\theta} - 1 \right) \sin\theta = 0,$$

when $\theta = \alpha$ or β.

In the azimuthal case it is usual for the map to extend to the pole, so we will take $\alpha = 0$, and find

$$A \tan \frac{\theta}{2} - B \cot \frac{\theta}{2} - 2 \cot \frac{\theta}{2} \log \sec \frac{\theta}{2} = 0,$$

when $\theta = 0$ or β. Now the limit to which $\cot \frac{\theta}{2} \log \sec \frac{\theta}{2}$ tends as $\theta \to 0$ is zero, and therefore we must have $B = 0$ and

$$A = 2 \cot^2 \frac{\beta}{2} \log \sec \frac{\beta}{2},$$

giving as the final expression for r in the azimuthal case

$$2 \left[\cot^2 \frac{\beta}{2} \log \sec \frac{\beta}{2} \tan \frac{\theta}{2} + \cot \frac{\theta}{2} \log \sec \frac{\theta}{2} \right].$$

As already indicated Airy took $A = 1$, which gives the true meridian scale at the pole, but does not make the square error a minimum. This minimum may be shewn to be

$$2 \sin^2 \frac{\beta}{2} - 8 \left(\log \sec \frac{\beta}{2} \right)^2 \cot^2 \frac{\beta}{2},$$

which, expanded to the first term, equals $\dfrac{\beta^6}{384}$, and is half that found for the orthomorphic case; nor is this first term altered if the total area of the map be made true by taking

$$A = 2 \left(\cos \frac{\beta}{2} - \cot^2 \frac{\beta}{2} \log \sec \frac{\beta}{2} \right).$$

Returning to the general conical case for a zone, we should find in the same way

$$A = \frac{\displaystyle\int_{\alpha}^{\beta} \tan^n \frac{\theta}{2} \, d\theta}{\tan^{2n} \dfrac{\beta}{2} - \tan^{2n} \dfrac{\alpha}{2}},$$

$$B = \frac{\tan^{2n} \dfrac{\alpha}{2} \displaystyle\int^{\beta} \tan^n \frac{\theta}{2} \, d\theta - \tan^{2n} \dfrac{\beta}{2} \displaystyle\int^{\alpha} \tan^n \frac{\theta}{2} \, d\theta}{\tan^{2n} \dfrac{\beta}{2} - \tan^{2n} \dfrac{\alpha}{2}},$$

and $\quad M = \cos \alpha - \cos \beta - 2n \dfrac{\left[\displaystyle\int_{\alpha}^{\beta} \tan^n \frac{\theta}{2} \, d\theta \right]^2}{\tan^{2n} \dfrac{\beta}{2} - \tan^{2n} \dfrac{\alpha}{2}},$

or with the notation previously adopted

$$M = 2 \sin \chi \sin \delta - 2n \frac{\left[\displaystyle\int_{\chi-\delta}^{\chi+\delta} \tan^n \frac{\theta}{2} \, d\theta \right]^2}{\displaystyle\int_{\chi-\delta}^{\chi+\delta} \frac{\tan^{2n} \dfrac{\theta}{2}}{\sin \theta} K_1 (2n, \theta) \, d\theta}.$$

Expanding numerator and denominator of this last fraction as in Chapter II, we find for the first

$$8n\delta^2 \tan^{2n} \frac{\chi}{2} \left[1 + \frac{n^2 - n\cos\chi}{\sin^2\chi} \frac{\delta^2}{2} \right.$$

$$+ \frac{\delta^4}{360\sin^4\chi}(16n^4 - 56n^3\cos\chi + 24n^2 + 52n^2\cos^2\chi$$

$$\left. - 30n\cos\chi - 6n\cos^3\chi) \right],$$

and for the second inverted

$$\frac{\sin\chi}{4n\delta \tan^{2n}\frac{\chi}{2}} \left[1 - \frac{4n^2 - 6n\cos\chi + 1 + \cos^2\chi}{\sin^2\chi} \frac{\delta^2}{6} \right.$$

$$\left. + \frac{\begin{Bmatrix}112n^4 - 240n^3\cos\chi + 140n^2\cos^2\chi - 40n^2 \\ + 90n\cos\chi - 30n\cos^3\chi + 7\cos^4\chi - 34\cos^2\chi - 5\end{Bmatrix}}{\sin^4\chi} \right] \frac{\delta^4}{360}.$$

Multiplying these two together, and putting $\frac{dM}{dn} = 0$ for a minimum we obtain

$$\frac{4(n - \cos\chi)}{\sin^2\chi}\frac{\delta^2}{6} + \frac{\delta^4}{360\sin^4\chi}(192n^3 - 288n^2\cos\chi$$

$$+ 104n\cos^2\chi - 72n + 80\cos\chi - 16\cos^3\chi)\ldots = 0.$$

Solving for n by successive approximation we obtain

$$n = \cos\chi\left(1 + \frac{\delta^2}{30}\right),$$

and substituting this in the expression for M we find

$$M = \frac{2\delta^5}{45}\sin\chi,$$

which agrees, as far as we have taken it, with the values obtained for the three projections already mentioned.

The expression for r will be seen to be the mean of two parts, $2A \tan^n \frac{\theta}{2}$, the orthomorphic part, and

$$2B \cot^n \frac{\theta}{2} + 2\cot^n \frac{\theta}{2} \int \tan^n \frac{\theta}{2}\, d\theta,$$

which satisfies the equation

$$\left(\frac{dr}{d\theta} - 1\right) = -\left(\frac{nr}{\sin\theta} - 1\right),$$

i.e. has its errors of scale along the meridians and parallels equal and opposite, and is thus approximately equal area; and each of these parts, if considered as the radius of a parallel will be found to require the same values of A and B as above, but will give a value of M double the absolute minimum, as we have already seen (p. 53) in the orthomorphic case.

Of these values A may be expanded in a manner similar to that applied to M: the numerator is

$$2\delta \tan^n \frac{\chi}{2} \left[1 + \frac{n^2 - n\cos\chi}{\sin^2\chi} \frac{\delta^2}{6} \cdots\right],$$

which, multiplied by the reciprocal of the denominator gives

$$\frac{\cot^n \frac{\chi}{2} \sin\chi}{2n} \left[1 + \frac{-3n^2 + 5n\cos\chi - 1 - \cos^2\chi}{\sin^2\chi} \frac{\delta^2}{6} \cdots\right],$$

or as far as terms in δ^2

$$A = \frac{\cot^n \frac{\chi}{2} \sin\chi}{2n} \left(1 - \frac{\delta^2}{6}\right).$$

On the other hand B, containing as it does merely the corrected integrals, must be left in its present form.

To find M and n for a map extending to the pole, we have

$$M = 1 - \cos\beta - 2n \left[\int_0^\beta \tan^n \frac{\theta}{2} d\theta\right]^2 \cot^{2n} \frac{\beta}{2},$$

and the integral may here be reduced by means of the formula

$$\int_0^\beta \tan^n \frac{\theta}{2} d\theta = \frac{2}{n+1} \tan^{n+1} \frac{\beta}{2} - \frac{2}{n+3} \tan^{n+3} \frac{\beta}{2} - \frac{2}{n+5} \tan^{n+5} \frac{\beta}{2} \cdots.$$

Substituting in M and putting $\dfrac{dM}{dn} = 0$, we find eventually

$$(1 - n^2)(3 + n)^4 (5 + n)^2$$
$$- (3 - n^2)(1 + n^2)(5 + n)^2 (3 + n)^2 2 \tan^2 \frac{\beta}{2}$$
$$+ [2(5 - n^2)(3 + n)^4 (1 + n)^2$$
$$+ (9 - n)^2 (1 + n)^4 (5 + n)^2] \tan^4 \frac{\beta}{2} \ldots = 0.$$

And solving by successive approximation

$$n = 1 - \tfrac{1}{2} \tan^2 \frac{\beta}{2} + \tfrac{4}{9} \tan^4 \frac{\beta}{2} \ldots$$
$$= 1 - \frac{\beta^2}{8} + \frac{\beta^4}{144},$$

giving $M = \dfrac{\beta^6}{1536}$.

Just as for the map of a zone we should find that the values of M in the orthomorphic and approximately equal area projections are double this, while that for Airy's projection is four times as great; also comparing with this the values obtained for the projections with meridian scale true, which previously we only considered in the case of a zone, as being the most likely to occur, we should find that if we take $n = 1 + a\beta^2$, then

$$M = \frac{\beta^6}{216}(1 + 18a + 108a^2),$$

giving $a = -\tfrac{1}{12}$ for minimum error—which also makes the total area true—and $M = \dfrac{\beta^6}{864}$, which is $\tfrac{16}{9}$ of the absolute minimum. So it appears that for maps extending to the pole the true meridian condition cannot be made to produce, as it does for a zone, a projection approximating to the minimum error; such maps, however, are of very infrequent occurrence except in the azimuthal case, with which we have already dealt.

II. Retro-azimuthal Projections

These projections possess the property converse to that of the azimuthal, in which the bearing of every point from the centre is true, and we thus have to search for the condition that the bearing of the centre from every point should be true. This property may readily be seen to belong to the oblique conical orthomorphic, for in that projection all angles, except at the centre, are preserved, and great circles through the centre become straight lines; hence the angle at any point P between the meridian and the great circle PC is equal to its projection, the angle between the projected meridian and the line pc. The only other projection of this class is the Mecca, which has been used for a map of Egypt and the Near East and is of obvious value to Mohammedans, being centred on that city.

To find the general condition, suppose p is the point whose polar co-ordinates referred to c are r, ω, and that ψ is the angle between cp and the projected meridian. Then

$$\tan \psi = -\, r\, \frac{d\omega}{dr},$$

where in the differentiation the longitude is kept constant,

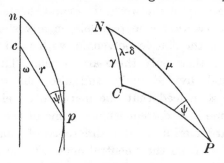

since this is only true as we move along a meridian. Then, if CNP is the corresponding spherical triangle, $\angle\, CPN = \psi$ and

$$\frac{1}{r}\frac{dr}{d\omega} = -\cot\psi = \frac{\sin\gamma\cos\mu\cos(\lambda - \delta) - \cos\gamma\sin\mu}{\sin\gamma\sin(\lambda - \delta)}.$$

The Mecca projection has for its meridians parallel lines at their true distances on the central parallel and hence

$$r = \sin \gamma \, (\lambda - \delta) \, \text{cosec} \, \omega.$$

Differentiating logarithmically we have

$$\frac{1}{r} \frac{dr}{d\omega} = - \cot \omega,$$

therefore $\omega = \psi$ and

$$\cot \omega = \frac{\cos \gamma \sin \mu - \sin \gamma \cos \mu \cos (\lambda - \delta)}{\sin \gamma \sin (\lambda - \delta)},$$

and for the rectangular co-ordinates

$$x = D \sin \gamma.$$

$$y = \frac{D \cos \gamma \sin \mu - \sin \gamma \cos \mu \cos D}{\sin D},$$

or reversing its direction

$$\sin l + \tfrac{1}{6} D^2 \sin l,$$

where $D = \lambda - \delta, \quad l = \gamma - \mu.$

III. The Projection for the International Map

This is a variation of the simple polyconic which has been adopted for a map of the world in several sheets. According to Hinks the two extreme parallels on any sheet are constructed, as in the simple polyconic, with radii tan $(\chi \pm \delta)$, but of the parallels only these are circular. The meridians are linear and divided truly, and also, the two extreme parallels are so placed that one meridian on either side of the central one is of correct length: some of the sheets have these standard meridians at a difference of longitude of 2° and some of 4° from their central ones. The co-ordinates of points on either of the extreme parallels, referred to the centres of those parallels, will be

$$x = \tan (\chi \pm \delta) \sin \{D \cos (\chi \pm \delta)\},$$
$$y = \tan (\chi \pm \delta) \cos \{D \cos (\chi \pm \delta)\},$$

as for the simple polyconic, but owing to a mistake they appear to have been calculated with the first factor a sine instead of a tangent. However, the theory is that points on these having the same longitude are joined to form the meridian, which is then divided truly. But before we can find expressions for the general co-ordinates we must first find the new distance between the extreme parallels, which is not as in the simple polyconic.

Let us suppose that the standard meridians are at a longitude difference ϕ from the central one and that the extreme parallels are at a distance $2b$ apart; and let x_1, y_1; x_2, y_2 be points on them and on one of the standard meridians, the co-ordinates being referred to axes through the point of intersec-

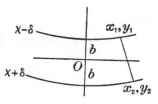

tion of the central parallel and the central meridian. Then

$$x_2 - x_1 = \tan(\chi + \delta) \sin\{\phi \cos(\chi + \delta)\}$$
$$- \tan(\chi - \delta) \sin\{\phi \cos(\chi - \delta)\}$$
$$= 2\delta\phi \cos\chi \ldots .$$

$$y_2 - y_1 = -2b + \tan(\chi + \delta) - \tan(\chi - \delta)$$
$$- \tan(\chi + \delta) \cos\{\phi \cos(\chi + \delta)\}$$
$$+ \tan(\chi - \delta) \cos\{\phi \cos(\chi - \delta)\}$$
$$= -2b + \delta\phi^2 \cos 2\chi \ldots .$$

And since $(x_2 - x_1)^2 + (y_2 - y_1)^2 = 4\delta^2$, we find

$$b^2 - b\phi^2\delta \cos 2\chi = \delta^2 (1 - \phi^2 \cos^2\chi),$$

and taking the square root we find

$$b = \delta \left(1 - \frac{\phi^2}{2} \sin^2\chi\right).$$

Now take a point $x, y,$ of latitude and longitude $\chi - l, D,$ and suppose x_1, y_1; x_2, y_2 are the points where its meridian cuts the extreme parallels. Then since the meridians are divided truly

$$x = \frac{(\delta + l)\,x_1 + (\delta - l)\,x_2}{2\delta} = x_1 + x_2 - \frac{l\,(x_2 - x_1)}{2\delta},$$

and expanding, we find

$$x_1 + x_2 = 2D \sin \chi \cos \delta \left(1 - \frac{D^2}{6} \{\cos^2\chi - (4\cos^2\chi - 1)\,\delta^2\} \dots \right),$$

$$x_1 - x_2 = 2D \cos \chi \sin \delta \left(1 - \frac{D^2}{6} \cos^2\chi \dots \right),$$

whence

$$x = D \sin \chi \cos \delta \left(1 - \frac{D^2}{6} \cos^2\chi \dots \right)$$

$$- Dl \cos \chi \left(1 - \frac{\delta^2}{6} \right) \left(1 - \frac{D^2}{6} \cos^2\chi \dots \right).$$

Similarly
$$y = \frac{y_1 + y_2}{2} - \frac{l\,(y_2 - y_1)}{2\delta},$$

$$y_1 + y_2 = \frac{D^2}{2} \sin 2\chi \cos 2\delta,$$

$$y_1 - y_2 = -\,2b + \frac{D^2}{2} \sin 2\delta \cos 2\chi,$$

and
$$y = l \left(1 - \frac{\phi^2}{2} \sin^2\chi \right) + \frac{D^2}{4} \sin 2\chi \cos 2\delta - \frac{lD^2}{2} \cos 2\chi.$$

IV. The Doubly Equidistant Projection

This projection, which has the property that the distances of all points from two given points are correct, was proposed by Col. Close at the meeting of the R.G.S. previously referred to on p. 61, and a map of the Atlantic Ocean constructed on it, appeared in the *Geographical Journal* for June 1921. It is of interest owing to its obvious value, for example, on a boat plying regularly between two ports, and also mathematically as it introduces the expansion of the arc of the great circle joining two points. We deal in a later chapter with the correction necessary if the ellipticity of the earth be regarded, and here we will confine ourselves to the case of the sphere.

The position of any point according to this projection can of course be found as the intersection of two arcs drawn with the two fixed points as centres and with radii

$$\cos^{-1}\{\cos\theta\cos\beta + \sin\theta\sin\beta\cos(\delta - \phi)\},$$

$$\cos^{-1}\{\cos\theta\cos\eta + \sin\theta\sin\eta\cos(\zeta - \phi)\},$$

where β, δ; η, ζ are the co-ordinates of the fixed points. Now while the expansion of these radii, or rather their squares, may be effected, attempts to find suitable expressions for the rectangular co-ordinates of the general point do not meet with any success, and it will be sufficient here to indicate the lines on which the arc A of the sphere may be expanded.

The direct application of MacLaurin's Theorem to the inverse cosine will be found to fail owing to the appearance in the denominator at each differentiation of $\sin A$, which vanishes when the differences of latitude and longitude are equated to zero. But we may circumvent this difficulty by using the expansions obtained for the oblique simple conical projection, which give

$$A\sin\phi = D\sin\gamma - lD\cos\gamma - \frac{l^2D}{3}\sin\gamma - \frac{D^3}{6}\sin\gamma\cos^2\gamma \ldots,$$

$$'A\cos\phi = l + \frac{D^2}{2}\sin\gamma\cos\gamma - \frac{lD^2}{6}\frac{\sin 3\gamma}{\sin\gamma} \ldots,$$

whence we obtain

$$A^2 = l^2 + D^2\sin^2\gamma - lD^2\sin\gamma\cos\gamma$$
$$- \frac{D^4}{12}\sin^2\gamma\cos^2\gamma - \frac{l^2D^2}{3}\sin^2\gamma \ldots,$$

for the square of the arc measured from a point of colatitude γ to any other point, the differences of latitude and longitude being l and D. This expression will enable the distances of particular points from the two fixed centres to be found, and then the points may be fixed by resection.

V. The Doubly Azimuthal and Rectilinear Projections

In the first of these projections, suggested by Mr A. R. Hinks, the azimuths of any point from two given points are correct. Supposing first of all one of the given points to be

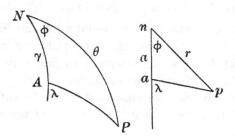

the pole and the other the point A, on the central meridian, colatitude γ, we find by considering the spherical triangle NAP, and the map triangle nap,

$$r = \frac{a}{\cos\phi\,(1 - \cos\gamma) + \sin\gamma\cot\theta},$$

where $a = na$, and if we take na for axis of y, positive downwards,

$$x = \frac{a\sin\phi}{\cos\phi\,(1 - \cos\gamma) + \sin\gamma\cot\phi},$$

$$y = \frac{a\cos\phi}{\cos\phi\,(1 - \cos\gamma) + \sin\gamma\cot\phi}.$$

The meridians are obviously the lines $x = y\tan\phi$, and the parallels will be found to be the curves

$$x^2\sin^2\gamma + y^2\{\sin^2\gamma - (1 - \cos\gamma)^2\tan^2\theta\}$$
$$- 2a\tan^2\theta\,(1 - \cos\gamma)\,y = a^2\tan^2\theta,$$

which will be ellipses or hyperbolae according as

$$\theta < \text{or} > \frac{\pi - \gamma}{2}.$$

More generally, let us suppose that instead of one of the fixed points being the pole, it is the point B (β, δ), while the other A is η, ζ. This amounts to applying the last projection obliquely; and if we take for axes the line ba, and the perpendicular to it, we have

$$x = \frac{r \cos (\phi - \nu)}{(1 - \cos \gamma) \cos (\phi - \nu) + \sin \gamma \cot \theta},$$

$$y = \frac{r \sin (\phi - \nu)}{(1 - \cos \gamma) \cos (\phi - \nu) + \sin \gamma \cot \theta},$$

where $ba = r$, $BA = \gamma$, and $\nu = \pi - \angle ABN$. These give

$$\cos \phi = \frac{\sin \gamma \cot \theta}{r - x (1 - \cos \gamma)} (x \cos \nu - y \sin \nu),$$

$$\sin \phi = \frac{\sin \gamma \cot \theta}{r - x (1 - \cos \gamma)} (x \sin \nu + y \cos \nu).$$

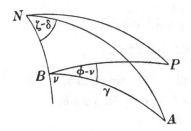
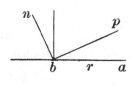

Substituting these in the equation of the meridian of an oblique projection, we obtain the line

$$y \{\sin \eta \cos \beta \cos (\lambda - \zeta) - \cos \eta \sin \beta \cos (\lambda - \delta)\}$$
$$+ x \{\sin \beta \sin (\lambda - \delta) - \sin \eta \sin (\lambda - \zeta)\} = r \sin \beta \sin (\lambda - \delta),$$

P being the point μ, λ.

Also, substituting for ϕ, θ and ν in the expression for y in terms of μ, λ, etc. we find

$$y = \frac{r}{1 - \cos \gamma} \cdot \frac{\left\{\begin{array}{l}\{\sin \eta \cos \beta \cos (\lambda - \zeta) - \sin \beta \cos \eta \sin (\lambda - \zeta)\} \sin \mu \\ \qquad + \sin \beta \sin \eta \sin (\zeta - \delta) \cos \mu\end{array}\right\}}{\left\{\begin{array}{l}\{\sin \eta \cos (\lambda - \zeta) + \cos \beta \cos (\lambda - \delta)\} \sin \mu \\ \qquad + (\cos \eta + \cos \beta) \cos \mu\end{array}\right\}}.$$

Putting $\mu = 0$ we have for the ordinate of the pole

$$\frac{r}{1 - \cos \gamma} \frac{\sin \beta \sin \eta \sin (\zeta - \delta)}{\cos \eta + \cos \beta},$$

and transferring the origin to the pole, keeping the axes parallel to their original directions, we find

$$y = \frac{r (1 + \cos \gamma)}{(1 - \cos \gamma)(\cos \eta + \cos \beta)}$$
$$\times \frac{\{\sin \eta \sin (\lambda - \zeta) - \sin \beta \sin (\lambda - \delta)\} \sin \mu}{\{\sin \eta \cos (\lambda - \zeta) + \sin \beta \cos (\lambda - \delta)\} \sin \mu + (\cos \eta + \cos \beta) \cos \mu},$$

and, using the equation of the meridian,

$$x = \frac{r (1 + \cos \gamma)}{(1 - \cos \gamma)(\cos \eta + \cos \beta)}$$
$$\times \frac{\{\sin \eta \cos \beta \cos (\lambda - \zeta) - \cos \eta \sin \beta \cos (\lambda - \delta)\} \sin \mu}{\{\sin \eta \cos (\lambda - \zeta) + \sin \beta \cos (\lambda - \delta)\} \sin \mu + (\cos \eta + \cos \beta) \cos \mu}.$$

These co-ordinates we may write in the shorter form

$$x = k \frac{a \sin \lambda + b \cos \lambda}{f \sin \lambda + g \cos \lambda + h \cot \mu},$$

$$y = k \frac{c \sin \lambda + d \cos \lambda}{f \sin \lambda + g \cos \lambda + h \cot \mu},$$

where
$$a = \sin \eta \cos \beta \sin \zeta - \cos \eta \sin \beta \sin \delta,$$
$$b = \sin \eta \cos \beta \cos \zeta - \cos \eta \sin \beta \cos \delta.$$
$$c = \sin \eta \cos \zeta - \sin \beta \cos \delta,$$
$$d = \sin \beta \sin \delta - \sin \eta \sin \zeta,$$
$$f = \sin \eta \cos \zeta + \sin \beta \cos \delta,$$
$$g = \sin \eta \sin \zeta + \sin \beta \sin \delta,$$
$$h = \cos \eta + \cos \beta,$$
$$k = \frac{r (1 + \cos \gamma)}{(1 - \cos \gamma)(\cos \eta + \cos \beta)}.$$

It will be seen that the central meridian $\lambda = 0$ is the line $y = \frac{d}{b} x$, and if we take this for axis of y we may reduce the expression for x to the form

$$\frac{A \sin \lambda}{f \sin \lambda + g \cos \lambda + h \cot \mu},$$

but the expression for y then contains coefficients of somewhat inconvenient form if expressed in terms of β, δ, η, ζ, and we will therefore leave the co-ordinates as above.

It will be noticed that the meridians both in the oblique and the normal projections are straight lines, shewing that great circles through any point are straight lines; and this is also true of the projection given by these equations whatever the values of the coefficients abc, etc., as may be seen by substituting in the equation of a great circle already obtained (p. 71). This therefore may be called the Rectilinear projection, and it will be seen to include as a particular case the gnomonic which, as a matter of fact, is the only one in which the scale errors are reasonably small.

The general expansions are:

$$x = \frac{k}{h} \left[\frac{b}{h} \mu + \frac{a}{h} \lambda\mu - \frac{bg}{h^2} \mu^2 - \frac{bf + ag}{h^2} \lambda\mu^2 \right.$$
$$\left. + \frac{b}{4h} \lambda^2\mu + \frac{b(h^2 - 3g^2)}{3h^3} \mu^3 \ldots \right],$$

$$y = \frac{k}{h} \left[\frac{d}{h} \mu + \frac{c}{h} \lambda\mu - \frac{dg}{h^2} \mu^2 - \frac{df + cg}{h^2} \lambda\mu^2 \right.$$
$$\left. + \frac{d}{4h} \lambda^2\mu + \frac{d(h^2 - 3g^2)}{3h^3} \mu^3 \ldots \right].$$

We here conclude, with one exception, the study of the equations giving the various projections and the nature of the parallels and meridians, and in the two chapters that follow we shall be dealing with measurements obtained from a map and the errors that arise owing to the representation of the earth on a plane according to those projections. We first consider errors of scale, that is of the ratios of infinitesimal parts, but as these errors vary from point to point we shall also have to study the differences between finite quantities measured on a map and their actual equivalents on the earth, and so again arises the question of the actual distance between two points on the earth and the expansion of the arc of a curve both on the sphere and the spheroid.

CHAPTER VII

THE DEFORMATION OF PROJECTIONS

The problem of the deformation at any point produced by
a projection really presents itself from two different points
of view, for we may use the study of it as a method of comparing one projection with another, or we may require to
find from measurements taken on a map the actual lengths
or angles to which those measurements correspond. For the
first problem we should require results to be given in terms
of quantities on the earth; for example, we might find the
alteration produced in an angle by the projection in terms
of that angle; while for the second we should require results
in terms of quantities measured on the map, and solve prob-
lems of the type "What is the actual angle to which a
measured one corresponds?"

Tissot, in his *Mémoire sur la Représentation des Surfaces et
les Projections des Cartes Géographiques*, makes a very complete study of the first problem, and from it produces his
method of finding the best projection for a given country,
of which some account is given later, and here we will follow
his method to a certain extent.

Any projection of one surface on to another, he says in
formulating his law of deformation, is equivalent to an infinite
number of orthogonal projections and variations from centres
of similitude; for every point, with certain exceptions, there
is a particular plane on to which the projection is made, and
a particular factor by which lengths are multiplied; and if
round any point a small circle be drawn it will be projected
into an ellipse, the study of which gives us the nature of the
deformation at that point. This ellipse, which in the particular case of orthomorphic projections becomes a circle, is
called the Indicatrix.

To prove the law we must assume that an angle of 180°
at a point projects into another angle of 180°, but this is not
always true, as at the pole of a conical projection, so there
may be exceptional points; however, with this assumption,
suppose XOX', OY be two lines at right angles which project

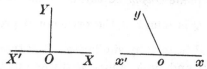

into xox', oy, not necessarily at right angles, and imagine a
right angle with its vertex at O moving from the position
$X'\hat{O}Y$ to $Y\hat{O}X$. The corresponding angle in the projected
figure first of all coincides with $x'oy$ and is acute, and finally
with yox and is obtuse. Somewhere between these two posi-
tions it must be a right angle. Hence there is at least one
pair of orthogonal lines OA, OB which project into an orthogonal pair oa, ob. This pair of lines through O are called the
principal tangents at O to the first surface.

If these tangents be drawn at every point they form a
graticule of orthogonal lines on the first surface which project
into a similar graticule on the second; and in the case of the
spheroid and the plane the tangents to the parallels and
meridians will be the principal tangents
in all cases in which the parallels and
meridians on the map are orthogonal,
but only in such cases. We will see
later how to find the direction of these
principal tangents in all cases.

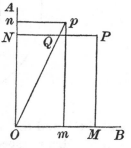

Now suppose the second surface is
applied to the first at the point O so
that oa, ob lie along OA, OB respec-
tively, and suppose that P is any point adjacent to O, and
p the projection of P. Then if Op cuts PN at Q,

$$\frac{QN}{pn} = \frac{ON}{On} = \frac{1}{a}, \qquad \frac{pn}{PN} = \frac{Om}{OM} = b,$$

where a and b are the scales in the two directions of the principal tangents. Hence

$$\frac{QN}{PN} = \frac{b}{a} \quad \text{and} \quad \frac{Op}{OQ} = a.$$

Thus if the plane $Onpm$ be turned about ON through an angle $\cos^{-1}\dfrac{b}{a}$, Q is found as the orthogonal projection of P, and p is formed from Q by similarity. If P trace out a small circle centre O and unit radius, p traces out the indicatrix at o, which is thus seen to be an ellipse, and is rather more conveniently formed by first increasing the radius of the circle to OR, $= a$, and then projecting on to a plane inclined to the original at $\cos^{-1}\dfrac{b}{a}$.

Now if OP and OP' are two orthogonal radii their projections Op, Op' will be conjugate semi-diameters of the ellipse,

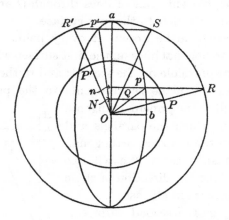

and if OP and OP' are elements of the arcs of the parallels and meridians respectively, Op, Op' will be elements of the arcs of the corresponding curves on the map; also if

$$Op' = h, \quad Op = k, \quad p\hat{O}p' = \epsilon,$$
$$a^2 + b^2 = h^2 + k^2, \quad \text{and} \quad ab = hk \sin \epsilon,$$

giving

$$a = \frac{(h^2 + k^2 + 2hk \sin \epsilon)^{\frac{1}{2}} + (h^2 + k^2 - 2hk \sin \epsilon)^{\frac{1}{2}}}{2},$$

$$b = \frac{(h^2 + k^2 + 2hk \sin \epsilon)^{\frac{1}{2}} - (h^2 + k^2 - 2hk \sin \epsilon)^{\frac{1}{2}}}{2}.$$

These equations are sufficient to give us the values of a and b, the scales along the principal tangents, for if the equations of the projection be given as relations between the map co-ordinates x, y and the colatitude and longitude θ and ϕ, we have, since $h = \frac{Op'}{OP'}$ and $k = \frac{Op}{OP}$, and the elements of arc of the meridian and parallel on the spheroid are $\rho \, d\theta$ and $\xi \, d\phi$,

$$h = \frac{1}{\rho} \left[\left(\frac{\partial x}{\partial \theta} \right)^2 + \left(\frac{\partial y}{\partial \theta} \right)^2 \right]^{\frac{1}{2}}, \quad k = \frac{1}{\xi} \left[\left(\frac{\partial x}{\partial \phi} \right)^2 + \left(\frac{\partial y}{\partial \phi} \right)^2 \right]^{\frac{1}{2}},$$

and

$$\cos \epsilon = \frac{1}{hk\rho\xi} \left[\frac{\partial x}{\partial \theta} \cdot \frac{\partial x}{\partial \phi} + \frac{\partial y}{\partial \theta} \cdot \frac{\partial y}{\partial \phi} \right],$$

whence

$$a^2 + b^2 = \frac{1}{\rho^2} \left[\left(\frac{\partial x}{\partial \theta} \right)^2 + \left(\frac{\partial y}{\partial \theta} \right)^2 \right] + \frac{1}{\xi^2} \left[\left(\frac{\partial x}{\partial \phi} \right)^2 + \left(\frac{\partial y}{\partial \phi} \right)^2 \right],$$

$$ab = \frac{1}{\rho\xi} \left[\frac{\partial x}{\partial \phi} \cdot \frac{\partial y}{\partial \theta} - \frac{\partial x}{\partial \theta} \cdot \frac{\partial y}{\partial \phi} \right],$$

though it is best as a rule to calculate h, k, ϵ first and find a and b from the original pair of equations.

In all conical projections $\epsilon = \frac{\pi}{2}$ and a is the greater and b the less of h and k. For example, in the azimuthal equal area for the sphere

$$x = 2 \sin \frac{\theta}{2} \sin \phi, \quad y = 2 \sin \frac{\theta}{2} \cos \phi,$$

$$h = \cos \frac{\theta}{2}, \quad k = \sec \frac{\theta}{2},$$

and thus

$$a = \sec \frac{\theta}{2}, \quad b = \cos \frac{\theta}{2}.$$

The ratio of the areas of the indicatrix and the circle gives the scale of areas, and therefore the equal area condition is $ab = 1$, or as we obtained before,

$$\frac{\partial x}{\partial \phi} \cdot \frac{\partial y}{\partial \theta} - \frac{\partial x}{\partial \theta} \cdot \frac{\partial y}{\partial \phi} = \rho \xi.$$

To find the direction of the axes of the indicatrix suppose OP' is the tangent to the meridian at O, and that

$$P'Oa = u, \quad p'Oa = v.$$

Then
$$a \cos u = h \cos v,$$

$$b \sin u = h \sin v,$$

whence
$$\sin u = \sqrt{\frac{a^2 - h^2}{a^2 - b^2}},$$

$$\sin v = \frac{b}{h} \sqrt{\frac{a^2 - h^2}{a^2 - b^2}},$$

and
$$\tan v = \frac{b}{a} \sqrt{\frac{a^2 - h^2}{h^2 - b^2}},$$

or in terms of h, k and ϵ,

$$\tan 2v = \frac{k^2 \sin 2\epsilon}{h^2 + k^2 \cos 2\epsilon}.$$

This gives the angle between the major axis of the indicatrix and the meridian on the map, and if we require to find the angle made with, for example, the axis of y on the map, we must subtract

$$\tan^{-1} \frac{\dfrac{\partial x}{\partial \theta}}{\dfrac{\partial y}{\partial \theta}},$$

i.e. the angle between the meridian and that axis. For example, in the Cassini projection for the sphere,

$$v = -\tfrac{1}{3} \phi \sin^3\theta \tan 2\theta.$$

For comparing one projection with another, having regard to the deformation of angles, we may conveniently study the values of the maximum alteration produced in an angle at any point.

From the triangles $OR'p'$, $Op'S$ we find

$$\sin (u - v) = \frac{a - b}{a + b} \sin (u + v).$$

Hence if the maximum value of $u - v$ be ω,

$$\sin \omega = \frac{a - b}{a + b}, \quad \text{and} \quad \tan \frac{\omega}{2} = \frac{\sqrt{a} - \sqrt{b}}{\sqrt{a} + \sqrt{b}}.$$

Suppose the values of u and v corresponding to this be u_0 and v_0, then

$$u_0 + v_0 = \frac{\pi}{2},$$

whence and since

$$\tan v_0 = \frac{b}{a} \tan u_0,$$

we have $\qquad \tan u_0 = \sqrt{\frac{a}{b}}, \quad \tan v_0 = \sqrt{\frac{b}{a}}.$

These results give the angles on the earth and the map in which the maximum alteration is produced by the projection, and the size of that alteration.

In terms of the meridian and parallel scales we have

$$\sin \omega = \sqrt{\frac{h^2 + k^2 - 2hk \sin \epsilon}{h^2 + k^2 + 2hk \sin \epsilon}}.$$

For example, in the Sanson-Flamsteed projection we find

$$\sin \omega = \frac{\phi \cos \theta}{\sqrt{4 + \phi^2 \cos^2 \theta}},$$

giving at the point

$$\phi = 45°, \quad \theta = 60°, \quad \sin \omega = \cdot 193,$$

while in the azimuthal equal area

$$\sin \omega = \frac{1 - \cos^2 \frac{\theta}{2}}{1 + \cos^2 \frac{\theta}{2}} = \cdot 143$$

for that point.

Again, in the simple conic with one standard parallel,

$$h = 1, \quad k = \frac{\sin \chi - l \cos \chi}{\sin (\chi - l)}, \quad \epsilon = \frac{\pi}{2},$$

giving
$$\sin \omega = \frac{l^2}{4} (1 + \tfrac{2}{3} l \cot \chi),$$

so that, from this point of view, the projection is suited to a map of some extent of longitude rather than of latitude; while on the other hand, for the simple polyconic

$$h = 1 + \frac{\phi^2}{2} \sin^2 \chi \dots, \quad k = 1,$$

$$\cos \epsilon = \phi^3 \sin^2 \chi \cos \chi \dots,$$

giving $\sin \omega$ approximately equal to $\frac{\phi^2}{4} \sin^2 \chi$, and the greatest possible error is least when the extent of longitude is small.

It must be borne in mind that this maximum alteration that we have calculated in the above cases is the maximum alteration possible in angles having one arm coincident with the first principal tangent, i.e. with that one corresponding to the major axis of the indicatrix, and therefore the maximum error possible for all angles round the point is 2ω. To calculate the alteration in a given angle u having one arm along the first principal tangent we have

$$\tan (u - u') = \frac{(a - b) \tan u}{a + b \tan^2 u}$$

$$= \frac{(a - b) \sin 2u}{a + b + (a - b) \cos 2u},$$

and if neither arm is in the necessary direction we must first find the alterations produced in the two parts into which the angle is divided by the first principal tangent, and then the total by addition or subtraction.

Turning from the variation of angles to that of scale in

any direction round a point, we should require to find, referring to the last figure, the value of Op' ($= r$). We have

$$r \cos v = a \cos u,$$

$$r \sin v = b \sin u,$$

giving $\qquad r^2 = a^2 \cos^2 u + b^2 \sin^2 u.$

Also $\qquad 2r \sin (u - v) = (a - b) \sin 2u;$

the maximum and minimum values of r correspond to the principal tangents and are a and b, and the direction in which the scale is true is

$$\tan^{-1} \sqrt{\frac{a^2 - 1}{1 - b^2}}.$$

If r_1, r_2 are the scales in two directions at right angles on the earth, then by the properties of conjugate diameters of the ellipse,

$$r_1{}^2 + r_2{}^2 = a^2 + b^2,$$

and $\qquad r_1 r_2 \cos \theta = ab,$

where θ is the alteration produced in the right angle by the projection.

Or again, suppose that the given direction makes an angle ϕ with the meridian, and that the first principal tangent divides ϕ into two parts u and v, the projections of ϕ, u, v being ϕ', u', v' respectively. Then

$$a \cos u = h \cos u',$$

$$a \cos v = r \cos v',$$

$$b \sin u = h \sin u',$$

$$b \sin v = r \sin v',$$

giving $\qquad hr \sin \phi' = ab \sin \phi,$

and since $\qquad hk \sin \epsilon = ab,$

$$r \sin \phi' = k \sin \epsilon \sin \phi.$$

Similarly $\qquad r \sin (\epsilon - \phi') = h \sin \epsilon \cos \phi.$

From these two equations we can at once determine r and ϕ' when ϕ is known. For instance, in the Sanson-Flamsteed,

$$h = (1 + D^2 \cos^2 \theta)^{\frac{1}{2}}, \quad k = 1, \quad \sin \epsilon = \frac{1}{(1 + D^2 \cos^2 \theta)^{\frac{1}{2}}},$$

for the point θ, D; and we find

$$r = [1 + 2D \cos \theta \sin \phi \cos \phi + D^2 \cos^2 \theta \cos^2 \phi]^{\frac{1}{2}}.$$

But all these results require knowledge of directions measured on the earth, our angles, for instance, having been measured from principal tangents or meridians or parallels; and it thus becomes necessary to find these directions first of all. A quantity often required is the true direction of the meridian at any point on the map, i.e. not the angle between the map meridian and a fixed line such as the y axis, but the angle on the earth corresponding to this. For in finding the true bearing of one point from another we measure the angle between the line joining them on the map and the y axis, supposed for the moment to be true north—and must then compound with it, by addition or subtraction, the angle made by the meridian at the point and that axis. This latter angle is called the Convergence of the Meridian, provided that the central meridian of the map is the axis of y; it is the angle between the meridian $\phi = $ constant, and the line on the earth corresponding to the line $x = $ constant. Now on this latter line we have

$$\frac{\partial x}{\partial \theta} d\theta + \frac{\partial x}{\partial \phi} d\phi = 0,$$

and therefore

$$\frac{d\phi}{d\theta} = -\frac{\dfrac{\partial x}{\partial \theta}}{\dfrac{\partial y}{\partial \theta}}.$$

Now the co-ordinates of any point on the spheroid are $\xi \cos \phi$, $\xi \sin \phi$, η; and the direction cosines of the tangent at this point corresponding to the line $x =$ constant on the map are thus

$$\frac{d\theta}{ds}\left(\rho\cos\theta\cos\phi + \xi\sin\phi\,\frac{\partial x}{\partial\theta}\Big/\frac{\partial x}{\partial\phi}\right)$$

$$\left[\text{since }\frac{d\xi}{d\theta}=\rho\cos\theta,\ \frac{\partial\eta}{\partial\theta}=\rho\sin\theta\right],$$

$$\frac{d\theta}{ds}\left(\rho\cos\theta\sin\phi - \xi\cos\phi\,\frac{\partial x}{\partial\theta}\Big/\frac{\partial x}{\partial\phi}\right),$$

$$-\frac{d\theta}{ds}\rho\sin\theta,$$

where ds is the element of arc,

$$=\left[\xi^2\left(\frac{\partial x}{\partial\theta}\Big/\frac{\partial x}{\partial\phi}\right)^2 + \rho^2\right]^{\frac12}d\theta.$$

Also the direction cosines of the meridian are $\cos\theta\cos\phi$, $\cos\theta\sin\phi$, $-\sin\theta$. Therefore if ω is the required convergence,

$$\cos\omega = \frac{\rho}{\left\{\xi^2\left(\frac{\partial x}{\partial\theta}\Big/\frac{\partial x}{\partial\phi}\right)^2 + \rho^2\right\}^{\frac12}},$$

giving
$$\tan\omega = \frac{\xi\,\dfrac{\partial x}{\partial\theta}}{\rho\,\dfrac{\partial x}{\partial\phi}} = \frac{\nu\sin\theta\,\dfrac{\partial x}{\partial\theta}}{\rho\,\dfrac{\partial x}{\partial\phi}}$$

$$= \frac{\sin\theta\,\dfrac{\partial x}{\partial\theta}}{\dfrac{\partial x}{\partial\phi}}(1 + 2e\sin^2\theta).$$

In a similar way we can find for the true bearing of the line $x = my + c$ at a given point, i.e. the angle the corresponding line on the earth makes with the meridian at the corresponding point, the value

$$\tan^{-1}\frac{\xi\left(\dfrac{\partial x}{\partial\theta} - m\dfrac{\partial y}{\partial\theta}\right)}{\rho\left(\dfrac{\partial x}{\partial\phi} - m\dfrac{\partial y}{\partial\phi}\right)}.$$

Or again, if we take two lines making with the axis of y angles whose tangents are m_1 and m_2, we should find for the value of the tangent of the angle on the earth represented by the angle between them,

$$\frac{\rho\xi\left(\dfrac{\partial x}{\partial \phi}\cdot\dfrac{\partial y}{\partial \theta}-\dfrac{\partial x}{\partial \theta}\cdot\dfrac{\partial y}{\partial \phi}\right)(m_2-m_1)}{\Delta_2 x-(m_1+m_2)\,\Delta_2 xy+m_1 m_2\Delta_2 y},$$

where

$$\Delta_2 x=\rho^2\left(\frac{\partial x}{\partial \phi}\right)^2+\xi^2\left(\frac{\partial x}{\partial \theta}\right)^2,$$

$$\Delta_2 xy=\rho^2\frac{\partial x}{\partial \phi}\cdot\frac{\partial y}{\partial \phi}+\xi^2\frac{\partial x}{\partial \theta}\cdot\frac{\partial y}{\partial \theta}.$$

Now this expression involves the angles made by the two arms of a given angle with the axis of y, just as, when using the indicatrix, we had to find the errors in two angles made with the first principal tangent. To avoid this necessity let us express the last result in terms of χ, the angle between the bisector of the given angle and the axis of y, and δ, half the given angle. This gives the angle on the earth as equal to

$$\tan^{-1}\frac{\rho\xi\left(\dfrac{\partial x}{\partial \phi}\cdot\dfrac{\partial y}{\partial \theta}-\dfrac{\partial x}{\partial \theta}\cdot\dfrac{\partial y}{\partial \phi}\right)}{\frac{1}{2}(\Delta_2 x-\Delta_2 y)\cos 2\chi-\Delta_2 xy\sin 2\chi+\frac{1}{2}(\Delta_2 x+\Delta_2 y)\cos 2\delta},$$

and we then have a convenient method of finding the directions of the lines on the map corresponding to the principal tangents; for the angles both on the earth and on the map are $\dfrac{\pi}{2}$, and therefore not only is $2\delta=\dfrac{\pi}{2}$, but the denominator of the above expression vanishes and we have

$$\tan 2\chi=\frac{\Delta_2 x-\Delta_2 y}{2\Delta_2 xy}.$$

A further application of this result is to the exceptions to the law of deformation. At the beginning of the chapter we assumed, following Tissot, that at any point a straight angle projected into a straight angle; but obviously the pole of a conical projection is an exception to this law, since there a straight angle becomes $n\pi$, and in other projections there may be similar points. Suppose then that the angle on the earth corresponding to an angle 2δ (not $= \pi$) at a given point is π. Then we must have

$$(1) \quad \rho\xi\left(\frac{\partial x}{\partial \phi}\cdot\frac{\partial y}{\partial \theta} - \frac{\partial x}{\partial \theta}\cdot\frac{\partial y}{\partial \phi}\right) = 0,$$

at the point in question, or

$$(2) \quad \text{One of } \frac{\partial x}{\partial \phi}, \ \frac{\partial x}{\partial \theta}, \ \frac{\partial y}{\partial \phi}, \ \frac{\partial y}{\partial \theta}$$

becomes infinite there. Possibility (2) will not be found to occur with any of the projections we have considered, at any rate at finite points; possibility (1) gives either $\xi = 0$, true at the poles, or

$$\frac{\partial x}{\partial \phi}\cdot\frac{\partial y}{\partial \theta} - \frac{\partial x}{\partial \theta}\cdot\frac{\partial y}{\partial \phi} = 0.$$

This last condition is obviously not fulfilled at any places other than the pole in the case of equal area projections, and the same may easily be shewn to be true for any conical projection. For the simple polyconic the condition gives

$$\cos(\phi\cos\theta)\sin^2\theta = 1,$$

and for any rectilinear projection

$$f\sin\lambda + g\cos\lambda + h\cot\mu = 0,$$

both yielding no result. It thus seems safe to assume Tissot's law for all points except the poles.

Finally let us investigate the scale at any point in a given direction $u\ (= \tan^{-1} m)$ with the axis of y. We find

$$\frac{d\phi}{d\theta} = -\frac{\dfrac{\partial x}{\partial \theta} - m\dfrac{\partial y}{\partial \theta}}{\dfrac{\partial x}{\partial \phi} - m\dfrac{\partial y}{\partial \phi}}.$$

Whence, finding first $\dfrac{dx}{d\theta}$ and $\dfrac{dy}{d\theta}$, we have

$$\left(\frac{dr}{d\theta}\right) = \frac{\left(\dfrac{\partial x}{\partial \phi} \cdot \dfrac{\partial y}{\partial \theta} - \dfrac{\partial x}{\partial \theta} \cdot \dfrac{\partial y}{\partial \phi}\right)(m^2 + 1)^{\frac{1}{2}}}{\dfrac{\partial x}{\partial \phi} - m \dfrac{\partial y}{\partial \phi}},$$

where dr is the element of length.

Then on the earth

$$\left(\frac{ds}{d\theta}\right)^2 = \rho^2 + \xi^2 \left(\frac{d\phi}{d\theta}\right)^2$$

$$= \frac{\Delta_2 x - 2m\,\Delta_2 xy + m^2 \Delta_2 y}{\left(\dfrac{\partial x}{\partial \phi} - m\,\dfrac{\partial y}{\partial \phi}\right)^2}.$$

Therefore the scale in this direction

$$= \frac{\left(\dfrac{\partial x}{\partial \phi} \cdot \dfrac{\partial y}{\partial \theta} - \dfrac{\partial x}{\partial \theta} \cdot \dfrac{\partial y}{\partial \phi}\right)(m^2 + 1)^{\frac{1}{2}}}{(\Delta_2 x - 2m\,\Delta_2 xy + m^2 \Delta_2 y)^{\frac{1}{2}}}$$

$$= \frac{\dfrac{\partial x}{\partial \phi} \cdot \dfrac{\partial y}{\partial \theta} - \dfrac{\partial x}{\partial \theta} \cdot \dfrac{\partial y}{\partial \phi}}{\left[\frac{1}{2}(\Delta_2 x + \Delta_2 y) - \Delta_2 xy \sin 2u + \frac{1}{2}(\Delta_2 x - \Delta_2 y)\cos 2u\right]^{\frac{1}{2}}}.$$

In an orthomorphic projection,

$$y + ix = f(\psi + i\phi),$$

whence
$$\frac{\partial x}{\partial \phi} = -\frac{\partial y}{\partial \psi} = -\frac{\xi}{\rho}\frac{\partial y}{\partial \theta},$$

$$\frac{\partial y}{\partial \phi} = +\frac{\partial x}{\partial \psi} = +\frac{\xi}{\rho}\frac{\partial x}{\partial \theta}.$$

Therefore $\Delta_2 x = \Delta_2 y$ and $\Delta_2 xy = 0$, and the expression for the scale is independent of u. Also we may find again the directions of the axes of the indicatrix, by finding the values of u that make r a maximum and a minimum. We find by differentiation that

$$(\Delta_2 x - \Delta_2 y) \sin 2u + 2\Delta_2 xy \cos 2u = 0,$$

giving $$\cot 2u = -\frac{\Delta_2 x - \Delta_2 y}{2\Delta_2 xy},$$

agreeing with the result obtained before for the direction of the bisector of the angle between the two directions since

$$u = \chi \pm \frac{\pi}{4}.$$

CHAPTER VIII

FINITE MEASUREMENTS

In the last chapter we found an expression for the true bearing of a line OP on the earth in terms of the angle between op on the map and the meridian through o; now we come to the calculation of the bearing of one point P from another O, which will in general be different from that if P is at a finite distance from O. Let us suppose for a moment that we are at O and want to get to P, and that we have a map in front of us; how shall we proceed? If we join op on the map and measure the angle between it and the meridian, and take that as a bearing for our march, we shall not reach P at all, supposing OP to be finite, unless the map is on Mercator's projection, for only in that case are angles preserved and meridians parallel. With all other projections either the angle on the earth corresponding to the map bearing varies with the latitude and longitude as we go along the line, or the map bearing itself varies from point to point, if the meridians converge, or both of these occur. If then we wish to follow the exact route corresponding to the straight line on the map we must be constantly changing our bearing; and this will generally be inconvenient, though we will give that route some consideration, and denote it by "the simple conic line," or "the Cassini line," according to the projection, or simply the map line.

The line of constant bearing from O to P is called the Loxodromic or Rhumb Line, and this is the route that is ordinarily used in navigation for short journeys; on the other hand long voyages, e.g. across the Atlantic, are made approximately on great circles (if the earth be regarded as a sphere) or geodesics, the lines of shortest distance from point to point, and the bearing is constantly being changed.

Let us see then what results can be obtained for the bearing and distance of P from O according to these three different routes, keeping in mind, however, that they are not always different, for the Mercator line is loxodromic and the gnomonic line a geodesic.

Suppose the two points are $x_1 y_1$, $x_2 y_2$, and that

$$\frac{x_2 - x_1}{y_2 - y_1} = \tan \alpha,$$

so that α is the map bearing measured from the axis of y. Then the differential equation of a map line, by the result of the last chapter, is

$$\frac{d\phi}{d\theta} = - \frac{\dfrac{\partial x}{\partial \theta} - \tan \alpha \dfrac{\partial y}{\partial \theta}}{\dfrac{\partial x}{\partial \phi} - \tan \alpha \dfrac{\partial y}{\partial \phi}}.$$

If this equation can be solved and we can find $\dfrac{d\phi}{d\theta}$ in terms of θ, we can then substitute in the expression for the arc

$$s = \int_{\theta_1}^{\theta_2} \sqrt{\rho^2 + \xi^2 \left(\frac{d\phi}{d\theta}\right)^2}\, d\theta.$$

In the case of a conical projection we find

$$\frac{d\phi}{d\theta} = - \frac{1}{nr} \frac{dr}{d\theta} \tan (n\phi - \alpha),$$

giving $kr = \operatorname{cosec} (n\phi - \alpha)$ for every point of the map line, and

$$\frac{d\phi}{d\theta} = - \frac{1}{nr} \frac{dr}{d\theta} \frac{1}{\sqrt{k^2 r^2 - 1}}.$$

k is a constant of integration, which, since $x_1 y_1$, $x_2 y_2$ lie on the map line, we find to be $\dfrac{d}{x_1 y_2 - x_2 y_1}$, where d is the map distance between the points.

In all orthomorphic cases

$$\frac{\xi}{nr} \frac{dr}{d\theta} = \rho,$$

and hence in the conical orthomorphic

$$s = \int_{\theta_1}^{\theta_2} \frac{kr\rho}{\sqrt{k^2r^2 - 1}}.$$

In the stereographic projection for the sphere s admits of complete integration in the form

$$\frac{4k}{\sqrt{4k^2 + 1}} \tan^{-1} \sqrt{\frac{k^2r^2 - 1}{4k^2 + 1}} = d \left(1 - \frac{\bar{x}^2 + \bar{y}^2}{4} - \frac{d^2}{48} \cdots \right),$$

where $\bar{x} = \frac{1}{2}(x_1 + x_2), \quad \bar{y} = \frac{1}{2}(y_1 + y_2).$

For the map line bearing of $x_2 y_2$ from $x_1 y_1$, we have from the last chapter

$$\tan \omega = - \frac{\dfrac{\xi}{\rho} \left\{ \left(\dfrac{\partial x}{\partial \theta}\right)_1 - \tan \alpha \left(\dfrac{\partial y}{\partial \theta_1}\right) \right\}}{\left(\dfrac{\partial x}{\partial \phi}\right)_1 - \tan \alpha \left(\dfrac{\partial y}{\partial \phi}\right)_1},$$

and in a conical projection

$$= \frac{\xi}{\rho} \frac{1}{nr} \frac{dr}{d\theta} \tan(\alpha - n\phi)$$

$$= \frac{\xi}{\rho} \tan \beta,$$

where β is the angle between the line op and the meridian, or what we have called the map bearing.

The only other case of interest in connection with map line bearings and distances is the Mercator, which we will use as an introduction to the consideration of loxodromic bearings and distances.

A loxodromic line cuts all meridians at the same angle, ω say, and so is defined by the differential equation

$$\frac{\xi}{\rho} \frac{d\phi}{d\theta} = \tan \omega,$$

or $\cot \omega \, d\phi = \dfrac{\rho}{\xi} d\theta = d\psi$ (cp. Chap. **IV**),

giving $\phi \cot \omega = \psi + $ a constant, and

$$\omega = \tan^{-1} \frac{\phi_2 - \phi_1}{\psi_2 - \psi_1},$$

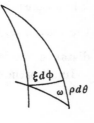

the loxodromic bearing required. The distance will be seen to be $\sec \omega \int_{\theta_1}^{\theta_2} \rho \, d\theta$. In the mercator we have

$$\omega = \tan^{-1} \frac{x_2 - x_1}{y_2 - y_1} = \alpha,$$

which we have already indicated to be the case.

Now we have to be able to calculate these quantities, viz. the bearing and distance, not in terms of the colatitude and longitude, as they are here given, but in terms of the map co-ordinates x and y, and of course the results will vary with the projection. The general method which we will give here will not be carried to its absolute conclusion, because there will occur so many different terms, all of which would not be present in the case of any particular projection, and which render the final results far too cumbersome to be of any value; but of course, in a particular case, when only some of the terms are present, the result will be easier to handle. In order to express our quantities in terms of x and y we must first find θ and ϕ in terms of them. This we will do by taking the general expressions

$$x = A\phi \sin \gamma - Bl\phi \cos \gamma - Cl^2\phi \cos \gamma \cot \gamma - Dl^2\phi \sin \gamma$$
$$- \frac{E\phi^3}{6} \sin \gamma \cos^2 \gamma,$$

$$y = A'l - B'l^2 \cot \gamma - \frac{C'\phi^2}{2} \cos \gamma \sin \gamma - \frac{D'l\phi^2}{2} \cos^2 \gamma$$
$$- \frac{E'l^3}{6},$$

where ϕ is the longitude, l the difference of latitude measured from the mid-parallel γ, and the coefficients are functions of γ, e and the depth of the map or distance between the standard parallels only. These will be found to include all the projections we have been dealing with, as far as terms of the third degree.

Now let these series be reversed, and we find

$$\phi \sin \gamma = ax + bxy \cot \gamma + cx^3 \cot^2 \gamma + (d \cot^2 \gamma + e)\, x^2 y,$$

$$l = a'y + b'x^2 \cot \gamma + c'y^2 \cot \gamma + d'x^2 y \cot^2 \gamma$$
$$+ (e' \cot^2 \gamma + f')\, y^3,$$

where $\quad a = \dfrac{1}{A}, \qquad b = \dfrac{B}{A^2 A'}, \qquad c = \dfrac{A'E + 3BC'}{6A^4 A'},$

$$d = \frac{B^2 A' + AA'C + ABB'}{A^3 A'^3}, \quad e = \frac{D}{A^2 A'^2},$$

$$a' = \frac{1}{A'}, \qquad b' = \frac{C'}{2A^2 A'}, \qquad c' = \frac{B'}{A'^3},$$

$$d' = \frac{2AB'C' + 2A'BC' + AA'D'}{2A^3 A'^3}, \quad e' = \frac{2B'^2}{A'^5}, \quad f' = \frac{E'}{6A'^4}.$$

Now $\qquad \psi_2 - \psi_1 = \displaystyle\int_{\theta_1}^{\theta_2} \operatorname{cosec} \theta \, (1 - 2e \sin^2 \theta)\, d\theta$

$$= 2\delta \operatorname{cosec} \chi \left[(1 - 2e \sin^2 \chi) \right.$$
$$\left. + \frac{\delta^2}{6}\, (2 \cot^2 \chi + 1 + 2e \sin^2 \chi) + \dots \right],$$

where $\qquad \theta_1 + \theta_2 = 2\chi, \quad \theta_2 - \theta_1 = 2\delta,$
and e is the ellipticity.

$$-\delta = a'v + 2 \cot \gamma \, (b'u\bar{x} + c'v\bar{y})$$
$$+ d' \cot^2 \gamma \, (\bar{x}^2 v + 2u\bar{x}\bar{y} + u^2 v) + e' \cot^2 \gamma v \, (3\bar{y}^2 + v^2) + \dots,$$

where $\bar{x}\bar{y}$ are as before (p. 114) and

$$u = \tfrac{1}{2}(x_2 - x_1), \quad v = \tfrac{1}{2}(y_2 - y_1),$$

$$\chi = \gamma - a'\bar{y} - \cot \gamma \, \{b'\,(\bar{x}^2 + u^2) + c'\,(\bar{y}^2 + v^2)\},$$

$$\sin \chi = \sin \gamma \left(1 - a'\bar{y} \cot \gamma - \{b'\,(\bar{x}^2 + u^2) \right.$$
$$\left. + c'\,(\bar{y}^2 + v^2)\} \cot^2 \gamma - \frac{a'^2 \bar{y}^2}{2} \right),$$

whence $\psi_2 - \psi_1$

$$= \frac{2a'v}{\sin \chi} \left(1 + 2 \cot \gamma \frac{(b' \tan \alpha \bar{x} + c'\bar{y})}{a'} \right) \{1 - 2e \sin^2 \gamma \,(1 - 2a'\bar{y} \cot \gamma)\},$$

and

$$(\phi_2 - \phi_1) \sin \gamma = 2au \left(1 + \frac{b}{a} \cot \gamma \, (\cot \alpha \, \bar{x} + \bar{y}) \right.$$

$$\left. + \, c \cot^2 \gamma \, (3\bar{x}^2 + u^2) + (d \cot^2 \gamma + e)(\bar{y}^2 + 2 \cot \alpha \, \bar{x}\bar{y} + v^2) \right),$$

where all the series are expanded as far as second degree terms only.

From these results we can find the expression for tan ω in any particular case, the series there being expanded as far as the first degree terms only, and it must be remembered that since the unit of length is the radius of the earth these terms will usually be small.

Having found tan ω we find sec ω and multiply by

$$\int_{\theta_1}^{\theta_2} \rho \, d\theta = \int_{\theta_1}^{\theta_2} \{1 + e \, (1 - 3 \sin^2 \theta)\} \, d\theta$$

$$= 2\delta \, \{1 + e \, (1 - 3 \sin^2 \chi) - e\delta^2 \cos 2\chi\}$$

$$= 2a'v \left\{ 1 + e \, (1 - 3 \sin^2 \gamma) \right.$$

$$\left. + \cot \gamma \left(\frac{2b'}{a'} \, \bar{x} \cot \alpha + \frac{c' + 6aa'e \sin^2 \gamma}{a'} \, \bar{y} \right) \right\},$$

omitting the negative sign.

It will be seen that the results are somewhat simplified in cases when

$$\frac{a}{a'} \, (1 + 2e \sin^2 \gamma) = 1,$$

a condition satisfied by the conical orthomorphic with one standard parallel, Mercator's and Gauss' Conformal for the spheroid, while the simple conic, conical orthomorphic and equal area, all with one standard parallel, Mercator's, Bonne's, the Simple Polyconic, Cassini's and Gauss' Conformal projections for the sphere, all have the coefficients in the expressions for x and y either ± 1 or 0, and the terms of the first degree in the expressions for the ratio of the distances vanish.

Following the above method results in any particular case may be found to any degree of accuracy required, and since the earth's equatorial and polar diameters are respectively 7926½ and 7899¼ miles, making $e = \cdot00344$, it will in most cases be possible to neglect the ellipticity. But in some projections results are interesting.

In the conical orthomorphic, with one standard parallel, the loxodromic lines are equiangular spirals, and we find

$$\tan \omega = \tan \alpha \{1 + 2\bar{x} \operatorname{cosec} 2\alpha \cot \gamma (1 - e \cos^2 \gamma)\}.$$

The expression for $\dfrac{s}{d}$ does not involve terms of the first degree, and may be taken as for the sphere.

In Mercator's projection $\omega = \alpha$, and

$$s = d \left\{1 - \left(\frac{\bar{y}^2}{2} + \frac{\bar{v}^2}{6}\right)(1 + 2e)\right\}.$$

Taking here $\bar{y} = \frac{1}{40}$, $\bar{v} = \frac{1}{80}$, corresponding roughly to 100 and 50 miles respectively, the difference between s and d is only about $\cdot034$ per cent., so that the results are for the most part quite small.

We now turn to the consideration of distances and bearings calculated from the geodesic, or shortest distance between two points on the sphere or spheroid. On such a curve we must have

$$\int (\xi^2 d\phi^2 + d\sigma^2)^{\frac{1}{2}} \text{ a minimum,}$$

i.e. $$\int (\xi^2 + \sigma'^2)^{\frac{1}{2}} d\phi \text{ a minimum,}$$

where $$\sigma' = \frac{d\sigma}{d\phi}.$$

Hence, by the Calculus of Variations, since ϕ is not explicitly present, we have for the first integral

$$(\xi^2 + \sigma'^2)^{\frac{1}{2}} = \frac{\sigma'^2}{(\xi^2 + \sigma'^2)^{\frac{1}{2}}} + c,$$

where c is a constant, or

$$\frac{\xi^2}{(\xi^2 + \sigma'^2)^{\frac{1}{2}}} = c \qquad \ldots\ldots(1).$$

Therefore if a geodesic cut a meridian at an angle ω, at a distance ξ from the axis, $\xi \sin \omega = c$; and we have to calculate c so that the geodesic passes through our two specified points.

From (1) we have

$$\xi^4 = c^2 \xi^2 + c^2 \left(\frac{d\sigma}{d\theta}\right)^2 \left(\frac{d\theta}{d\phi}\right)^2,$$

whence

$$\left(\frac{d\phi}{d\theta}\right)^2 = \frac{c^2 (1-e)^4}{(1 - 2e\cos^2\theta)^3 \xi^2 (\xi^2 - c^2)},$$

and

$$\phi = -\int \frac{c(1-e)^2 \sqrt{1+z^2}}{\sqrt{1 + (1-e)^2 z^2} \sqrt{1 - c^2 \{1 + (1-e)^2 z^2\}}}$$

(where $z = \cot\theta$)

$$= \cos^{-1} \frac{c \cot\theta}{\sqrt{1 - c^2}} + e\left[\frac{c \cot\theta}{(1 - c^2 \mathrm{cosec}^2\theta)^{\frac{1}{2}}} - c \cos^{-1}\frac{\cos\theta}{\sqrt{1-c^2}}\right] + \beta,$$

by expanding the expression under the integral sign to the first power of e and then integrating.

The constants c and β must be chosen to make the geodesic pass through θ_1, ϕ_1, and θ_2, ϕ_2, and we find c by the method of successive approximation.

Suppose $c = c_0 (1 + be)$. Then, considering first of all only terms without e, we have

$$\phi_2 - \phi_1 = \cos^{-1} \frac{c_0 \cot\theta_2}{\sqrt{1 - c_0^2}} - \cos^{-1} \frac{c_0 \cot\theta_1}{\sqrt{1 - c_0^2}},$$

whence

$$c_0 = \frac{\sin\theta_1 \sin\theta_2 \sin(\phi_2 - \phi_1)}{\sin A},$$

where A is the arc of the sphere, as is otherwise evident from spherical trigonometry. Now we must substitute $c = c_0(1 + be)$ in the expression for $\phi_2 - \phi_1$, and equate to zero the coefficient of e. This gives

$$b = (1 - c_0^2) \left[1 - \frac{\cos^{-1}\dfrac{\cos\theta_2}{\sqrt{1 - c_0^2}} - \cos^{-1}\dfrac{\cot\theta_1}{\sqrt{1 - c_0^2}}}{\dfrac{\cos\theta_2}{\sqrt{\sin^2\theta_2 - c_0^2}} - \dfrac{\cos\theta_1}{\sqrt{\sin^2\theta_1 - c_0^2}}} \right],$$

and we must remember that, since b is the coefficient of e, and we are only including the first power of e, we may in the calculation of b suppose the earth a sphere.

Now

$$\sqrt{\sin^2 \theta_1 - c_0{}^2} = -\frac{\sin \theta_1}{\sin A}\{\sin \theta_1 \cos \theta_2 - \cos \theta_1 \sin \theta_2 \cos (\phi_2 - \phi_1)\}$$

and

$$\sqrt{\sin^2 \theta_2 - c_0{}^2} = \frac{\sin \theta_2}{\sin A}\{\cos \theta_1 \sin \theta_2 - \sin \theta_1 \cos \theta_2 \cos (\phi_2 - \phi_1)\},$$

the negative sign appearing in the first expression since the quantity in brackets, as may be seen from the spherical triangle, is

$$-\sin \theta_2 \sin (\phi_2 - \phi_1) \cot \lambda_1.$$

Hence we find the numerator of the second term of b to be A and the denominator

$$\frac{\sin^3 A\,(1 - c_0{}^2)}{\begin{Bmatrix}\sin \theta_1 \sin \theta_2 \{\sin \theta_1 \cos \theta_2 - \cos \theta_1 \sin \theta_2 \cos (\phi_2 - \phi_1)\} \\ \{\sin \theta_2 \cos \theta_1 - \cos \theta_2 \sin \theta_1 \cos (\phi_2 - \phi_1)\}\end{Bmatrix}},$$

giving

$$b = (1 - c_0{}^2) - \frac{A}{\sin^3 A} \sin \theta_1 \sin \theta_2 \{\sin \theta_1 \cos \theta_2 - \cos \theta_1 \sin \theta_2$$
$$\times \cos (\phi_2 - \phi_1)\}\{\sin \theta_2 \cos \theta_1 - \cos \theta_2 \sin \theta_1 \cos (\phi_2 - \phi_1)\}.$$

Then, b having been found, the angle between the geodesic and the meridian at $\theta_1 \phi_1$, or the bearing of $\theta_2 \phi_2$ from $\theta_1 \phi_1$ is given by

$$\sin \omega = \frac{c_0}{\xi_1}(1 + be) = \frac{c_0}{\sin \theta_1}\{1 + e\,(b - \cos^2 \theta_1)\}$$

$$= \sin \lambda_1 \left[1 + e\,\frac{\sin \theta_1}{\sin^2 A}\{\sin \theta_1 \cos \theta_2 - \cos \theta_1 \sin \theta_2 \cos (\phi_2 - \phi_1)\}\right.$$

$$\times \left(\sin \theta_1 \{\sin \theta_1 \cos \theta_2 - \cos \theta_1 \sin \theta_2 \cos (\phi_2 - \phi_1)\}\right.$$

$$\left.\left.-\frac{A \sin \theta_2}{\sin A}\{\sin \theta_2 \cos \theta_1 - \cos \theta_2 \sin \theta_1 \cos (\phi_2 - \phi_1)\}\right)\right],$$

or

$$\sin \lambda_1 \left\{1 + e \sin \theta_1 \cos \lambda_1 \left(\sin \theta_1 \cos \lambda_2 + \frac{A}{\sin A}\sin \theta_2 \cos \lambda_2\right)\right\}.$$

For the length of the arc of the geodesic we have

$$s = \int_{\theta_1}^{\theta_2} \left[\xi^2 \left(\frac{d\phi}{d\theta} \right)^2 + \left(\frac{d\sigma}{d\theta} \right)^2 \right]^{\frac{1}{2}} d\theta$$

$$= (1 - e)^2 \int_{\theta_1}^{\theta_2} \frac{\sin \theta \, d\theta}{(1 - 2e \cos^2 \theta)^{\frac{3}{2}} \{ \sin^2 \theta - c^2 (1 - 2e \cos^2 \theta) \}^{\frac{1}{2}}}.$$

In evaluating this result to the first power of e, note that

$$\int \frac{\sin \theta \, d\theta}{(\sin^2 \theta - c^2)^{\frac{1}{2}}} = \cos^{-1} \frac{\cos \theta}{\sqrt{1 - c^2}},$$

$$(1 - c^2) \int \frac{\sin \theta \, d\theta}{(\sin^2 \theta - c^2)^{\frac{3}{2}}} = - \frac{\cos \theta}{\sqrt{\sin^2 \theta - c^2}},$$

and

$$\int \frac{\sin^3 \theta \, d\theta}{(\sin^2 \theta - c^2)^{\frac{1}{2}}} = \frac{1 + c^2}{2} \cos^{-1} \frac{\cos \theta}{\sqrt{1 - c^2}} - \frac{\cos \theta \sqrt{\sin^2 \theta - c^2}}{2}.$$

Employing these we find

$$s_0 = \int_{\theta_1}^{\theta_2} \frac{\sin \theta \, d\theta}{(\sin^2 \theta - c_0^2)^{\frac{1}{2}}} = \left[\cos^{-1} \frac{\cos \theta}{\sqrt{1 - c_0^2}} \right]_{\theta_1}^{\theta_2} = A,$$

$$\left(\frac{ds}{de} \right)_0 = \left[- \frac{1 + c^2}{2} \cos^{-1} \frac{\cos \theta}{\sqrt{1 - c_0^2}} \right.$$

$$\left. + c_0^2 \left(1 - \frac{e^2}{1 - c_0^2} \right) \frac{\cos \theta}{\sqrt{\sin^2 \theta - c_0^2}} + \tfrac{3}{2} \cos \theta \sqrt{\sin^2 \theta - c_0^2} \right]_{\theta_1}^{\theta_2}.$$

The second term of this last becomes, when we put in the limits, $c_0^2 A$, and the third term

$$\frac{3}{2 \sin A} \{ (\sin^2 \theta_1 + \sin^2 \theta_2) \cos A - 2 \sin \theta_1 \sin \theta_2 \cos (\phi_2 - \phi_1) \}.$$

Thus the final expression is

$$s = A \left[1 - e \left\{ \frac{1 - c_0^2}{2} - \frac{3}{2A \sin A} \{ (\sin^2 \theta_1 + \sin^2 \theta_2) \cos A \right. \right.$$

$$\left. \left. - 2 \sin \theta_1 \sin \theta_2 \cos (\phi_2 - \phi_1) \} \right\} \right].$$

Our next problem is to express $A s$ and ω in terms of the means and semi-differences of the map co-ordinates, and to

begin with let us consider the case of the sphere. We have
already seen how to expand the arc A in terms of the co-
ordinates of the two extremities (p. 93), and the only difference here is that we express x and y in terms of χ and δ,
equal to $\dfrac{\gamma + \mu}{2}$ and $\dfrac{\gamma - \mu}{2}$ respectively, i.e.

$$x = A\,\frac{\sin(\chi - \delta)\sin D}{\sin A},$$

$$y = A\,\frac{\cos(\chi - \delta) - \cos A \cos(\chi + \delta)}{\sin A \sin(\chi + \delta)},$$

where

$$\cos A = \cos(\chi - \delta)\cos(\chi + \delta) + \sin(\chi - \delta)\sin(\chi + \delta)\cos D$$

and
$$D = \phi_2 - \phi_1.$$

Expanding these expressions, squaring and adding, we find
$$A^2 = 4\delta^2 + D^2 \sin^2\chi + D^2\delta^2\left(\tfrac{2}{3}\sin^2\chi - 1\right)$$

$$- \frac{D^4}{12}\sin^2\chi\cos^2\chi + 2D^2\delta^3\sin\chi\cos\chi - \frac{D^4}{3}\delta\sin^3\chi\cos\chi$$

as far as terms of the fifth degree.

Also substituting in the expression for s,
$$\theta_1 = \chi - \delta, \quad \theta_2 = \chi + \delta,$$
we find for the arc of the spheroid

$$s = A\left[1 - \frac{e}{2A^2}\left\{4\delta^2\left(1 - 3\cos 2\chi\right) - 2D^2\sin^2\chi\cos^2\chi\right.\right.$$

$$+ \tfrac{8}{3}\delta^4\cos 2\chi + \frac{D^4}{6}\sin^2\chi\cos^4\chi$$

$$\left.\left. + D^2\delta^2\left(2 - \tfrac{4}{3}\sin^2\chi - \tfrac{4}{3}\sin^4\chi\right)\ldots\right\}\right].$$

The method we adopt is first to calculate the arc A of the
sphere in terms of $\bar{x}$, $\bar{y}$, u and v; and then s. Also having
found A, we can then find $\sin\lambda_1$, $\sin\lambda_2$ from the equations

$$\sin\lambda_1 = \frac{\sin\theta_2\sin D}{\sin A}, \quad \sin\lambda_2 = \frac{\sin\theta_1\sin D}{\sin A}.$$

Hence we have $\cos \lambda_1$ and $\cos \lambda_2$, and from these $\sin \omega$; or we may find ω more conveniently from the equation

$$\tan \omega = \tan \lambda_1 \left[1 + e \left\{ \sin^2 \theta_1 + \frac{A}{\sin A} \sin \theta_1 \sin \theta_2 \frac{\cos \lambda_2}{\cos \lambda_1} \right\} \right].$$

Let us take Mercator's projection as an example. Here

$$\phi = x, \quad l = y(1 + 2e) - \frac{y^3}{6}(1 + 10e),$$

$$4\delta^2 = 4v^2 \left\{ 1 + 4e - \left(\bar{y}^2 + \frac{v^2}{3} \right)(1 + 12e) \right\},$$

$$D = 2u, \quad \sin^2 \chi = 1 - \bar{y}^2 (1 + 4e).$$

From these we find

$$A = d \left[1 + 2e \cos^2 \alpha - \left(\frac{\bar{y}^2}{2} + \frac{v^2}{6} \right) \{1 + 4e(1 + 2\cos^2 \alpha)\} \right].$$

For the coefficient of e in the expression in the brackets for s, we neglect e and find

$$\frac{4}{d^2} \left(2v^2 - 2v^2 \bar{y}^2 - u^2 \bar{y}^2 - \frac{v^4}{3} - \frac{u^2 v^2}{3} \right)$$

$$= 2 \cos^2 \alpha - \bar{y}^2 (1 + \cos^2 \alpha) - \frac{v^2}{3},$$

and combining these two, we find

$$s = d \left(1 - \frac{\bar{y}^2}{2} \{1 + 2e(1 + 3\cos^2 \alpha)\} - \frac{v^2}{6} \{1 + 2e(1 + 4\cos^2 \alpha)\} \right).$$

For ω we find that

$$\sin \lambda_1 = \sin \alpha \left[1 + v\bar{y} + \frac{v^2}{3} + e \left\{ 2v\bar{y}(2 - \cos^2 \alpha) \right. \right.$$

$$\left. \left. + 5\bar{y}^2 \cos^2 \alpha + \tfrac{8}{3} u^2 \cos^2 \alpha - \frac{v^2}{3}(4 - 11\cos^2 \alpha) \right\} \right],$$

and when we come to calculate $\cos \lambda_1$ and $\cos \lambda_2$ we may neglect e, since they only appear in its coefficient. The final result obtained is

$$\tan \omega = \tan \alpha \left[1 + \left(v\bar{y} + \frac{v^2}{3} \right) \sec^2 \alpha \right.$$

$$\left. + e \left\{ 6v\bar{y} \tan^2 \alpha + 3\bar{y}^2 + \frac{10u^2}{3} - \frac{v^2}{3}(4 \sec^2 \alpha - 7) \right\} \right].$$

Results for the sphere are naturally much less complicated, and distances as far as terms of the second degree are for the most part very nearly the same as when calculated loxodromically.

Some results are given here for comparison:

Loxodromic

Simple Conic (one S.P.)

$$\begin{cases} A = d \left\{ 1 - \left(\dfrac{\bar{y}^2}{2} + \dfrac{v^2}{6} \right) \sin^2 \alpha + \dfrac{u^2 \cot^2 \gamma}{6} \ldots \right\} \\ \tan \omega = \tan \alpha \left(1 + 2\bar{x} \operatorname{cosec} 2\alpha \cot \gamma \ldots \right) \end{cases}$$

Conical Orthomorphic (one S.P.)

$$\begin{cases} A = d \left\{ 1 - \left(\dfrac{\bar{y}^2}{2} + \dfrac{v^2}{6} \right) + \dfrac{u^2 \cot^2 \gamma}{6} \ldots \right\} \\ \text{Same as simple conic to first degree} \end{cases}$$

Sanson-Flamsteed

$$\begin{cases} A = d \left(1 + \bar{x}\bar{y} \sin \alpha \cos \alpha + \dfrac{v^2}{3} \sin^2 \alpha \right) \\ \tan \omega = \tan \alpha \left(1 + \bar{x}\bar{y} + \dfrac{v^2}{3} \right) \end{cases}$$

Geodesic

Simple Conic (one S.P.)

$$\begin{cases} A = d \left\{ 1 - \left(\dfrac{\bar{y}^2}{2} + \dfrac{v^2}{6} \right) \sin^2 \alpha \right\} \\ \tan \omega = \tan \alpha \left\{ 1 + (2\bar{x} \operatorname{cosec} 2\alpha + v^2 \sec^2 \alpha) \cot \gamma \ldots \right\} \end{cases}$$

Conical Orthomorphic (one S.P.)

$$\begin{cases} A = d \left\{ 1 - \left(\dfrac{\bar{y}^2}{2} + \dfrac{v^2}{6} \right) \ldots \right\} \\ \text{Same as simple conic to first degree} \end{cases}$$

Sanson-Flamsteed

$$\begin{cases} A = d \left(1 + \bar{x}\bar{y} \sin \alpha \cos \alpha + \dfrac{v^2}{3} \sin^2 \alpha \right) \\ \tan \omega = \tan \alpha \left(1 + \bar{x}\bar{y} \cot \alpha + v\bar{y} \sec^2 \alpha + \dfrac{v^2}{3} (2 + \tan^2 \alpha) \right) \end{cases}$$

It will be noted that for the Mercator and Sanson-Flamsteed, for the sphere, the loxodromic and geodesic distances are the same, as far as terms of the second degree, while the two bearings are different.

The process converse to the above is worthy of some attention, though the method is fairly evident, viz. given the actual distance and bearing of $x_2 y_2$ from a known point $x_1 y_1$, to find expressions for $x_2 y_2$.

Suppose $x_1 y_1$ has colatitude $\delta\ (=\gamma - L)$ which will be known in terms of $x_1 y_1$ and the coefficients of the projection: then we have the expressions

$$A^2 = L^2 + D^2 \sin^2 \delta - LD^2 \sin \delta \cos \delta - \frac{D^4}{12} \sin^2 \delta \cos^2 \delta$$

$$- \frac{L^2 D^2}{3} \sin^2 \delta \dots \quad \text{(cp. p. 93)},$$

$$\sin \omega = \frac{\sin (\delta + L) \sin D}{\sin A},$$

where L and D are the differences of latitude and longitude between the two points and A and ω the given geodesic distance and bearing on the earth, neglecting the ellipticity. From the second equation

$$D \sin \delta = A \sin \omega \left\{ 1 - L \cot \delta + \frac{L^2}{2} (2 \cot^2 \delta + 1) \right.$$

$$\left. - \left(1 - \frac{\sin^2 \omega}{\sin^2 \delta}\right) \frac{A^2}{6} \dots \right\},$$

and if we substitute this in the first we have an equation in L which can be solved in ascending powers of A by successive approximation, the result being

$$L = A \cos \omega + \tfrac{3}{2} A^2 \sin^2 \omega \cot \delta + A^3 \{\tfrac{1}{2} \sin^2 \omega \cos \omega$$

$$- 2 \cot^2 \delta \sin \omega \tan \omega (1 - \tfrac{5}{3} \sin^2 \omega)\},$$

from which we also obtain

$$D \sin \delta = A \sin \omega \left(1 - A \cot \delta \cos \omega + \frac{A^2}{3}\right.$$

$$\left. \times \{(3 - 7 \sin^2 \omega) \cot^2 \delta + \cos^2 \omega\}\right).$$

If γ is the colatitude of the mid-parallel of the projection,

$$D \sin \gamma = D \sin (\delta + L)$$

$$= D \sin \delta \left(1 + L \cot \delta - \frac{L^2}{2} \dots\right),$$

and we are now in a position to apply the equations of the projection to find x_2 and y_2. For example, in the Mercator

$$x_2 = x_1 + A \sin \omega \left(1 - A y_1 \cos \omega + \frac{y_1^2}{2} \right),$$

$$y_2 = y_1 + A \cos \omega \left(1 + \tfrac{3}{2} A \sin \omega \tan \omega + \frac{y_1^2}{2} \right.$$
$$\left. + \frac{A y_1}{2} \cos \omega + \frac{A^2 \cos^2 \omega}{6} \right).$$

This method would apply equally well to the spheroid, though naturally the expressions are more complicated.

The only other question that arises in this connection is the expansion of Cassini's projection for 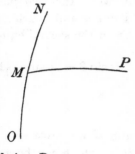 the spheroid. We saw in the case of the sphere that the projection could be obtained by expanding the lengths of the arcs PM, MO of the two great circles ON, the central meridian, and PM at right angles to it. Hence in the spheroid we must first draw the geodesic PM. Let the colatitude of M be α, and that of P, θ, the difference of longitude being D.

Then since $P\hat{M}O = 90°$, we have the equation

$$1 = \frac{\sin \alpha \sin \theta}{\sin A} \left\{ 1 + e \sin \alpha \, (\sin \alpha \cos \theta - \cos \alpha \sin \theta \cos D) \right.$$

$$\times \left(\sin \alpha \, (\sin \alpha \cos \theta - \cos \alpha \sin \theta \cos D) \right.$$
$$\left. \left. - \frac{A \sin \theta}{\sin A} (\sin \theta \cos \alpha - \cos \theta \sin \alpha \cos D) \right) \right\}.$$

Squaring, transposing and dividing by
$$\sin \alpha \cos \theta - \cos \alpha \sin \theta \cos D,$$
we find
$$\tan \alpha = \tan \theta \cos D + 2e \, \frac{\sin^2 \theta \sin^2 D}{\sin^2 A} \left\{ \sin \alpha \, (\tan \alpha - \tan \theta \cos D) \right.$$
$$\left. - \frac{A \sin \theta}{\sin A} (\tan \theta - \tan \alpha \cos D) \right\}.$$

Now in the coefficient of e we may regard the earth as a sphere, and so we find

$$\tan \alpha = \tan \theta \cos D \left\{ 1 - 2eD^2 \sin^2 \theta \left(1 - \frac{D^2 \cos 2\theta}{3} \right) \right\};$$

then by Maclaurin's theorem,

$$\alpha = \theta - \frac{D^2}{2} \sin \theta \cos \theta \, (1 + 4e \sin^2 \theta).$$

Therefore

$$MN = \left(1 - \frac{e}{2} \right) \alpha + \frac{3e}{4} \sin 2\alpha$$

$$= \left(1 - \frac{e}{2} \right) \theta + \frac{3e}{4} \sin 2\theta - \frac{D^2}{2} \sin \theta \cos \theta \, (1 + e + e \sin^2 \theta),$$

and $\qquad y = M + \dfrac{D^2}{2} \sin \theta \cos \theta \, (1 + e + e \sin^2 \theta),$

where M is the meridian distance of P from the equator.

Now applying the expression for the arc of the sphere we find

$$A = D \sin \theta \left(1 - \frac{D^2}{6} \cos^2 \theta \ldots \right),$$

and then for the arc of the spheroid

$$x = PM$$

$$= D \sin \theta \, (1 + e \cos^2 \theta) - \frac{D^3}{6} \sin \theta \cos^2 \theta \, (1 - 2e \cos^2 \theta).$$

CHAPTER IX

THE BEST PROJECTION FOR A GIVEN COUNTRY

USE OF RECTANGULAR CO-ORDINATES. THE CONFORMAL EXPANSION

In Chapter VII we used the theory of the indicatrix to compare one projection with another; now we are to use it, or at any rate part of it, to find the projection best suited to any given country, expanded after the method of Tissot. We shall seek for expressions for x and y which first of all make the scales h and k differ from unity by quantities as small as possible and the angle ϵ as near $\frac{\pi}{2}$ as possible, producing ω of the third order of small quantities, and then make the scale error on the boundary of the map as small as possible. Tissot uses for his co-ordinates the arc of the meridian s, measured from the mid-parallel, and the arc of the mid-parallel t, but we will keep to our usual co-ordinates $l \, (= \chi - \mu$, where χ is the colatitude of the centre), and D the difference of longitude.

Tissot's method is to deal separately with the terms of the different degrees. First of all the origin may be so chosen that no terms independent of l and D occur in the expressions for x and y, and if the tangent to the projection of the central meridian be made the axis of y, the first degree term in l will be missing from x and that in D from y.

Now

$$h = \frac{1}{\rho} \left[\left(\frac{\partial x}{\partial l} \right)^2 + \left(\frac{\partial y}{\partial l} \right)^2 \right]^{\frac{1}{2}}, \quad k = \frac{1}{\xi} \left[\left(\frac{\partial x}{\partial D} \right)^2 + \left(\frac{\partial y}{\partial D} \right)^2 \right]^{\frac{1}{2}},$$

and therefore, to make h and k approximate to unity our expressions must begin

$$x = \xi D, \quad y = \rho l.$$

Now $\left(\dfrac{\partial x}{\partial l}\right)^2$ does not introduce into h any term of the first degree, consequently $\left(\dfrac{\partial y}{\partial l}\right)^2$ must not; therefore y can contain no term in D^2 and none in lD except that already contained in ξD.

Now let us suppose that $x = \xi D + al^2$, $y = \rho l + bD^2$, and let it be borne in mind that as ρ is the same for two adjacent points on the meridian it is to be regarded as constant in the expansion.

Then

$$\cos \epsilon = \frac{1}{hk\xi\rho}\{(- D\rho \cos \mu + 2al)\,\xi + 2b\rho D\}$$

$$= \frac{1}{\rho_0\xi_0}\{D\,(2b - \xi_0 \cos \chi)\} + \frac{2al}{\rho_0}\,...,$$

where $\rho_0\xi_0$ are the values at the centre. Therefore to make the terms of the first degree in $\cos \epsilon$ vanish, we must take

$$a = 0, \quad b = \tfrac{1}{2}\xi_0 \cos \chi.$$

So we may assume the expansions

$$x = \xi D + \frac{El^3}{3} + Fl^2D \sin \chi + GlD^2 \sin^2\chi + \frac{H}{3}\,D^3 \sin^3\chi,$$

$$y = \rho l + \tfrac{1}{2}\xi_0 \cos \chi D^2 + \frac{E'l^3}{3} + F'l^2D \sin \chi + G'lD^2 \sin^2\chi$$
$$+ \frac{H'}{3}\,D^3 \sin^3\chi,$$

in which the introduction of the powers of $\sin \chi$ will be found to simplify the relations between the coefficients.

From these we have, as far as second degree terms,

$$h = 1 + \frac{E'l^2}{\rho_0} + \frac{2F'lD \sin \chi}{\rho_0} + \left(\frac{G' \sin^2\chi}{\rho_0} + \frac{\cos^2\chi}{2}\right)D^2,$$

$$k = 1 + \frac{Fl^2}{\xi_0}\sin \chi + \frac{2GlD \sin^2\chi}{\xi_0} + \left(\frac{H \sin^3\chi}{\xi_0} + \frac{\cos^2\chi}{2}\right)D^2,$$

$$\cos \epsilon = \frac{1}{\xi_0\rho_0}\left[(\xi_0 E + \rho_0 F' \sin \chi)\,l^2 + 2\left\{\xi_0 F \sin \chi + \rho_0 G' \sin^2\chi\right.\right.$$
$$\left.\left. - \frac{\rho}{2}(\xi_0 \sin \chi - \rho_0 \cos^2\chi)\right\}lD + (\xi_0 G + \rho_0 H' \sin \chi)D^2 \sin^2\chi\right].$$

Tissot then makes his projection approximately orthomorphic by taking

$$\xi_0 E' = \rho_0 \sin \chi F, \quad \xi_0 F' = \rho_0 \sin \chi G, \quad \xi_0 G' = \rho_0 \sin \chi H,$$

$$\xi_0 E + \rho_0 \sin \chi F'' = 0, \quad \xi_0 G + \rho_0 \sin \chi H' = 0,$$

$$\xi_0 \sin \chi F + \rho_0 \sin^2 \chi G' = \tfrac{1}{2}\rho_0 (\xi_0 \sin \chi - \rho_0 \cos^2 \chi),$$

and therefore obtains the expansions

$$x = \xi D + \frac{E l^3}{3} + F l^2 D \sin \chi - \frac{\xi_0^2 E}{\rho_0^2} l D^2 + \frac{H}{3} D^3 \sin^3\chi,$$

$$y = \rho l + \tfrac{1}{2}\xi_0 \cos \chi D + \frac{\rho_0 \sin \chi}{3\xi_0} l^3 - \frac{\xi_0 E}{\rho_0} l^2 D$$

$$+ \frac{\rho_0}{\xi_0} H \sin^3\chi \, l D^2 + \frac{\xi_0^3}{3\rho_0^3} D^3,$$

where

$$\xi_0^2 \sin \chi F + \rho_0^2 \sin^3 \chi H = \tfrac{1}{2}\rho_0 \xi_0 (\xi_0 \sin \chi - \rho_0 \cos^2 \chi).$$

This projection will have $h - 1$, $k - 1$ of the second order, $h - k$ and $\cos \epsilon$ of the third order, and therefore ω of the third order, for

$$\sin \omega = \frac{a - b}{a + b} \quad \text{and} \quad (a - b)^2 = (h - k)^2 \cos^2 \frac{\epsilon'}{2} + (h + k)^2 \sin^2 \frac{\epsilon'}{2},$$

where $\epsilon' = \dfrac{\pi}{2} - \epsilon$, and is therefore of the third order. Our problem now is to find the values of E, F and H which will make the scale error on the boundary of the map as small as possible. This scale error is approximately the same in all directions and is therefore

$$\frac{1}{\rho} \frac{\partial y}{\partial l} - 1 = \frac{\sin \chi}{\xi_0} F l^2 - \frac{2\xi_0 E}{\rho_0^2} l D + \frac{\sin^3 \chi}{\xi_0} H D^2$$

$$= \frac{\sin \chi}{\xi_0 \rho_0^2} F y^2 - \frac{2E}{\rho_0^3} x y + \frac{\sin^3\chi}{\xi_0^3} H x^2 \cdots,$$

by substituting $l = \dfrac{y}{\rho_0}$, $D = \dfrac{x}{\xi_0}$ from above. This expresses the scale error in terms of the map co-ordinates, and since

$$\frac{\sin\chi}{\xi_0\rho_0{}^2}F + \frac{\sin^3\chi}{\xi_0{}^3} = \frac{1}{2\xi_0{}^2\rho_0}\left(\xi_0\sin\chi - \rho_0\cos^2\chi\right)$$

$$< \frac{\sin\chi}{2\xi_0\rho_0}, \text{ i.e. } < \frac{1}{2\rho_0\nu_0},$$

where ν_0 is the grand normal at the centre, it follows that the maximum scale error I can be expressed in the form

$$I = \frac{1}{\rho_0\nu_0}\left[(\tfrac{1}{2} - F)\,x^2 - \frac{2\nu_0 E}{\rho_0{}^2}\,xy + Fy^2\right].$$

This shews that the line of equal error is an ellipse whose axes are inclined to the axes of co-ordinates at an angle α given by

$$\tan 2\alpha = \frac{\nu_0 E}{\rho_0{}^2\,(F - \tfrac{1}{4})},$$

and whose equation referred to principal axes is

$$I = \frac{1}{4\rho_0\nu_0}\{(1 + p)\,x^2 + (1 - p)\,y^2\},$$

where $\qquad p = -E\operatorname{cosec}2\alpha = -(F - \tfrac{1}{4})\sec 2\alpha,$

and if the ratio of the axes be r,

$$p = \frac{r^2 - 1}{r^2 + 1}.$$

Now if the radius of this ellipse inclined at $45°$ be R,

$$I = \frac{R^2}{4\rho_0\nu_0},$$

and we must therefore seek for the ellipse which surrounds the country and fits the shape of it as nearly as possible so as to make the radius R and therefore I a minimum. This latter quantity can be further reduced to one-half by the introduction of a scale constant

$$A = 1 - \frac{R^2}{8\rho_0\nu_0},$$

which produces an error at the centre equal and opposite to that on the boundary. Tissot's method is to take sheets of tracing paper and on each sheet draw a number of similar concentric ellipses, such that the ratio of the principal axes varies from sheet to sheet by tenths from 0 to 1; then to find by trial which shape of ellipse best fits the given country, mapped by plotting $x = \xi D$, $y = \rho l$, and also to measure the angle between the axis of x and the major axis of the ellipse in its best position. We thus find the best values of p and α, and therefore of E, F and H, and so are in a position to complete the expansions of x and y.

Arising out of this method of Tissot's are two matters demanding attention; first the study of curves on the sphere, since what we have been seeking by means of tracing paper is the equation of the ellipse that most nearly surrounds the country; and secondly the fact that Tissot has arrived, by an independent method of dealing in turn with each of the terms of the expansions, at a result that is true of all conformal projections, and can be used to determine the constants of such projections. Let us take the first problem first, and seek for some analytical method of finding the best ellipse surrounding a country.

To this end we must introduce rectangular co-ordinates u and v for a point on the sphere, instead of the polar coordinates θ and ϕ.

Suppose N is the pole or origin, NA the zero meridian and NB the meridian at right angles to it. These we take for axes. From any point P let great circles PU, PV be drawn at right angles to NA, NB respectively. Then P is defined by the length $NU = u$ and $NV = v$. These are the rectangular co-ordinates of P and we shall see

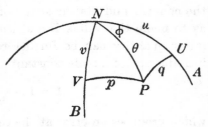

that in many ways their tangents are analogous to ordinary

plane rectangular co-ordinates. But before we set down any results let us emphasize the fact that the properties of parallel lines in a plane have no exact counterpart in the theory of great circles on a sphere. For example, the angles at N, U and V are right angles but not so that at P; nor does $NV = PU$ nor $NU = PV$. We must therefore apply special symbols p and q to PV and PU. These we may term the "back co-ordinates" of P, but it must be remembered that they are not the rectangular co-ordinates of N referred to axes through P.

The following results are derived immediately from the ordinary formulae of spherical trigonometry:

$$\tan u = \tan \theta \cos \phi, \qquad \sin p = \sin \theta \cos \phi = \sin u \cos q,$$
$$\tan v = \tan \theta \sin \phi, \qquad \sin q = \sin \theta \sin \phi = \sin v \cos p,$$
$$\cot \phi = \cot q \sin u, \qquad \tan p = \tan u \cos v,$$
$$\tan \phi = \cot p \sin v, \qquad \tan q = \tan v \cos u.$$

Now transforming the equation of the great circle through two points γ, δ; η, ζ or u_1, v_1; u_2, v_2 (see p. 71), from polars to rectangular co-ordinates, we have

$$\frac{\tan u - \tan u_1}{\tan u_2 - \tan u_1} = \frac{\tan v - \tan v_1}{\tan v_2 - \tan v_1},$$

except when either u or v is equal to $\dfrac{\pi}{2}$. This shews that any equation of the first degree in $\tan u$ and $\tan v$ represents a great circle, and conversely. And here we may mention, just in passing, that the use of the equation of a great circle in this form facilitates the finding of the intersection of two great circles in order to determine the pole of an oblique conical projection.

It now becomes evident that the simplest curve other than a great circle has an equation of the second degree in $\tan u$ and $\tan v$, since it will be cut by a great circle in two points only. Such an equation has five constants, and such a curve is therefore determined by five points. Suppose then that five points are chosen at the extremities of the given country, and their latitudes and longitudes and therefore their rectangular co-ordinates known; we can then find the equation of a second degree curve passing through them, of the form

$$a \tan^2 u + 2h \tan u \tan v + b \tan^2 v + 2g \tan u$$
$$+ 2f \tan v + c = 0.$$

Next we must find the centre, γ, ζ in polar, or ξ, η in the rectangular co-ordinates, and $\tan \xi$, $\tan \eta$ will be found from the equations

$$a \tan \xi + h \tan \eta + g - \tan \xi (g \tan \xi + f \tan \eta + c) = 0,$$
$$h \tan \xi + b \tan \eta + f - \tan \eta (g \tan \xi + f \tan \eta + c) = 0;$$

for as for a plane conic section the polar of the point ξ, η with respect to the curve is

$$(a \tan \xi + h \tan \eta + g) \tan u + (h \tan \xi + b \tan \eta + f) \tan v$$
$$+ g \tan \xi + f \tan \eta + c = 0,$$

and if ξ, η is the centre its polar with respect to the curve is its polar (or equator)

$$\tan \xi \tan u + \tan \eta \tan v + 1 = 0.$$

A comparison of these two equations leads to the pair above, from which cubic equations in $\tan \xi$ and $\tan \eta$ can be found. The roots of these give ξ and η and therefore γ and ζ.

Now turn the axes through an angle $\left(\dfrac{\pi}{2} - \zeta\right)$ in the negative direction, by means of the relations

$$\tan u = \tan u' \sin \zeta + \tan v' \cos \zeta,$$
$$\tan v = \tan u' \cos \zeta - \tan v' \sin \zeta,$$

where u', v' are the new current co-ordinates. This makes the new axis of v pass through the centre, and if we transfer the origin to the centre, by means of the relations

$$\tan u' = \frac{\tan u'' \sec \gamma}{1 - \tan v'' \tan \gamma},$$

$$\tan v' = \frac{\tan v'' + \tan \gamma}{1 - \tan v'' \tan \gamma},$$

we obtain the equation of the curve in the form

$$P \tan^2 u + 2Q \tan u \tan v + R \tan^2 v = 1.$$

Next, to find the expansions of
u and v in terms of l and D we
have

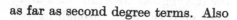

$$\tan u = \tan \theta \cos \phi$$
$$= \frac{\sin D \sin \mu}{\cos \mu \cos \gamma + \sin \mu \sin \gamma \cos D},$$

and $u = \dfrac{D \sin \mu}{\cos l} \cdots$

$$= D \sin \gamma + Dl \cos \gamma,$$

as far as second degree terms. Also

$$\tan v = \tan \theta \sin \phi$$
$$= \cot \gamma - \frac{\cos \mu}{\sin \gamma \, (\cos \mu \cos \gamma + \sin \mu \sin \gamma \cos D)},$$

giving

$$v = l - \frac{\cos \mu}{\sin \gamma} \frac{D^2}{2} \cdots.$$

Thus, as far as terms of the second degree, the equation of
the curve is

$$PD^2 \sin^2 \gamma + 2QlD \sin \gamma + Rl^2 = 1.$$

Now P, Q and R are known in terms of the original coefficients
of the curve, and therefore in terms of the five chosen points;
they thus suffice to give us the values of the coefficients E
and F in our expansion of the projection in terms of those
points, by virtue of the relations

$$F = \frac{R}{2 \, (P + R)}, \quad E = - \frac{Q}{2 \, (P + R)},$$

by comparing the above equation with that on p. 130.

An interesting simplification of this projection is when the ellipse becomes a circle, $Q = 0$ and $P = R$ giving $E = 0$, $F = \frac{1}{4}$, and for the expansions

$$x = D \sin \gamma - Dl \cos \gamma - \frac{Dl^2}{4} \sin \gamma - \frac{D^3}{12} \sin \gamma \, (2 - 3 \sin^2 \gamma),$$

$$y = l + \frac{D^2}{2} \sin \gamma \cos \gamma + \frac{l^3}{12} - \frac{lD^2}{4} \, (2 - 3 \sin^2 \gamma),$$

which agree with the expansions for the oblique stereographic found from the relation $y + ix = 2 \tan^{-1} \frac{f - \gamma}{2}$ of p. 66. This gives added weight to the supposition that this projection, the stereographic, is the best for a country of equal extent of latitude and longitude.

Now the above use of the rectangular co-ordinates u and v suggests that we might have used them all through instead of θ and ϕ, and it is worth while spending, before we close, a little time on one or two aspects of their use. First of all the question of the alteration of the degree of a curve by a projection is most easily solved in this way. The gnomonic projection $x = \tan v$, $y = \tan u$ is evidently one that does not alter the degree of a curve, and this, as we have already seen, is a special case of the doubly azimuthal

$$x = k \, \frac{a \tan v + b \tan u}{f \tan v + g \tan u + h},$$

$$y = k \, \frac{c \tan v + d \tan u}{f \tan v + g \tan u + h},$$

or $$\tan u = \frac{k \, (ay - cx)}{k \, (ad - bc) - (fd - gc) \, x - (ag - fb) \, y},$$

$$\tan v = \frac{k \, (dx - by)}{k \, (ad - bc) - (fd - gc) \, x - (ag - fb) \, y},$$

which will include all cases of such projections.

Then the Cassini projection is given very simply by the relations $x = q$, $y = u$. Finally, to find the orthomorphic relation involving these co-ordinates, we must first find the element of arc in terms of them. Suppose two great circles be drawn from adjacent points M and M' on the axis of u, both at right angles to that axis. They will meet at a point K such that

$$MK = M'K = \frac{\pi}{2}$$

and

$$M\hat{K}M' = MM' = \delta u.$$

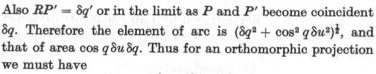

Let P be a point on MK, P' an adjacent point on $M'K$ and PR ($= \delta u'$) at right angles to $M'K$. Then

$$\frac{\sin PR}{\sin M\hat{K}M'} = \sin PK = \cos q,$$

$$\therefore \quad \delta u' = \cos q\, \delta u.$$

Also $RP' = \delta q'$ or in the limit as P and P' become coincident δq. Therefore the element of arc is $(\delta q^2 + \cos^2 q\, \delta u^2)^{\frac{1}{2}}$, and that of area $\cos q\, \delta u\, \delta q$. Thus for an orthomorphic projection we must have

$$x + iy = f(z + iu),$$

where

$$z = \int \sec q\, dq = \tfrac{1}{2} \log \frac{1 + \sin q}{1 - \sin q};$$

and taking for the function a constant of proportion k we have

$$x = \tfrac{1}{2}k \log \frac{1 + \sin q}{1 - \sin q}, \quad y = ku,$$

which will be seen to be the Gauss Conformal. Nor does it seem likely that other functions would lead to any projections of importance that could not be expressed as simply in terms of the colatitude and longitude; and it is only when dealing with curves on the sphere that we can use rectangular co-ordinates with more advantage than the others.

Now let us turn to the second matter arising out of Tissot's results. The expression for the scale error obtained on p. 131, viz. $I = \dfrac{1}{4\rho_0 \nu_0} \{(1 + p)\, x^2 + (1 - p)\, y^2\}$, which we will shew to be true for any conformal projection, brings us to the most recent developments in our subject, made by Commandant Laborde (April 1928) for a projection for Madagascar.

In an article in the *Geographical Journal* of October 1930 Young points out that some of Laborde's results are not really new, but are contained partly in the works of Gauss (1822) and in Lagrange's conformal projection for the spheroid. This latter and the conical orthomorphic, normal or oblique, are, as we shall see, really special cases of a more general conformal projection which may be determined in succession from the general equation

$$(1 + I)\left(\frac{\partial^2 I}{\partial x^2} + \frac{\partial^2 I}{\partial y^2}\right) - \left[\left(\frac{\partial I}{\partial x}\right)^2 + \left(\frac{\partial I}{\partial y}\right)^2\right] = \frac{1}{\rho \nu},$$

which is the equation obtained by Laborde in his pamphlet as something new, though it can easily be seen to be the same as

$$-\frac{1}{2m^2}\left[\frac{\partial^2 \log m^2}{\partial x^2} + \frac{\partial^2 \log m^2}{\partial y^2}\right] = \frac{1}{\rho \nu},$$

where m, the reciprocal of the scale, $= \dfrac{1}{1 + I}$, a fact pointed out by Young, who quotes from the works of Liouville where the result is attributed to Gauss.

To prove it, we have that if ds is the element of arc $ds^2 = m^2 (dx^2 + dy^2)$ and therefore $m^2 = \xi^2 J$, where J is the Jacobian

$$\begin{vmatrix} \dfrac{\partial \phi}{\partial x} & \dfrac{\partial \phi}{\partial y} \\[2mm] \dfrac{\partial \psi}{\partial x} & \dfrac{\partial \psi}{\partial y} \end{vmatrix}, \text{ or } \left(\frac{\partial \phi}{\partial x}\right)^2 + \left(\frac{\partial \psi}{\partial x}\right)^2,$$

since $\qquad\qquad \psi + i\phi = f(y + ix),$

giving $\qquad\quad \dfrac{\partial \phi}{\partial x} = \dfrac{\partial \psi}{\partial y}, \quad \dfrac{\partial \psi}{\partial x} = -\dfrac{\partial \phi}{\partial y},$

$$\frac{\partial^2 \phi}{\partial x^2} = -\frac{\partial^2 \phi}{\partial y^2}, \quad \frac{\partial^2 \psi}{\partial x^2} = -\frac{\partial^2 \psi}{\partial y^2},$$

$$\frac{\partial^3 \phi}{\partial x^3} = -\frac{\partial^3 \psi}{\partial x^3}, \quad \frac{\partial^3 \psi}{\partial x^3} = \frac{\partial^3 \phi}{\partial x^3}.$$

From these we obtain

$$\frac{\partial^2 J}{\partial x^2} + \frac{\partial^2 J}{\partial y^2} = 0, \quad \frac{\partial^2 \log \xi^2}{\partial x^2} + \frac{\partial^2 \log \xi^2}{\partial y^2} = J \frac{\partial^2 \log \xi^2}{\partial^2 \psi^2},$$

and therefore

$$-\frac{1}{2m^2} \left(\frac{\partial^2 \log m^2}{\partial x^2} + \frac{\partial^2 \log m^2}{\partial y^2} \right) = -\frac{1}{2\xi^2} \frac{\partial^2 \log \xi^2}{\partial \psi^2} = \frac{1}{\rho \nu}.$$

This equation, in Laborde's form above, may be solved by a series of terms which, if the co-ordinates be measured from the point where the scale is true and also a minimum, begins

$$I = \frac{1}{4\rho_0 \nu_0} \{(1 + p) x^2 + (1 - p) y^2\} + C (x^3 - 3xy^2)$$

$$+ \frac{\cos \chi (\xi_0 - \rho_0 \sin \chi)}{12\xi_0^2 \rho_0^2} \{3 (1 + q) x^2 y + (1 - q) y^3\} \dots.$$

Let us consider the first two terms first. This is Tissot's result and as before we find the best value of p by making a line of equal error coincide with the ellipse that best fits the country. This gives $p = \dfrac{r^2 - 1}{r^2 + 1}$, where r is the ratio of the axes.

Young prefers Airy's method of minimum error which may be conveniently applied to find a different value of p. If a scale constant $1 - a$ be introduced making

$$I = -a + \frac{1}{4\rho_0 \nu_0} \{(1 + p) x^2 + (1 - p) y^2\},$$

and $2I^2$ be integrated over a rectangular map whose sides are $2X$, $2Y$, we find that to make the total square error M a minimum,

$$p = \frac{r^4 - 1}{r^4 + 1}, \quad a = \frac{Y^2}{6\rho_0 \nu_0} \frac{(r^2 + 1)}{r^4 + 1}, \quad M = \frac{4Y^5 \sin \chi}{45\rho_0^2 \nu_0^2 (r^4 + 1)},$$

if we bear in mind the fact that in Chapter I, p. 8, we divided the original integral by the longitude, i.e. by $\dfrac{2X}{\sin \chi}$, to get M. Here $r = \dfrac{Y}{X}$. This agrees with the result obtained for the Mercator projection for which $r = 0$, and for the Gauss conformal, $r = \infty$, by interchanging X and Y and inverting r; also if $r = 1$, for a map of equal extent in either direction M is reduced to the minimum possible for a conical projection.

In just the same way, if we include the terms of the third degree in I and apply the condition of minimum error, we find $C = 0$ and

$$q = \frac{5r^4 - 21}{5r^4 - 21r^2 + 21},$$

while M is reduced by an amount

$$\frac{Y^7 \cos^2 \chi \, (\xi_0 - \rho_0 \sin \chi)^2}{24 \sin \chi \cdot \rho_0{}^4 \cdot \xi_0{}^4 \, (5r^4 - 21r^2 + 21)}.$$

Leaving the method of determining the constants let us consider the general projection. Working further on Laborde's method Young has obtained in hyperbolic functions complete formulae for the most general conformal projection, on which he read a paper at the General Assembly of the International Union of Geodesy and Geophysics at Stockholm in August 1930, but as his results are not yet published we can here do no more than discuss a little further the solution in series.

The conformal projections so far considered are successive attempts at the general one. If we take the normal conical projection we can by a suitable choice of A and n make I zero at the centre, and get rid of its first degree term. But our constants are then exhausted and the constant p which appears in the expression for I is determined already by the others and so cannot be chosen either to make I a minimum on the boundary or M a minimum over the map. Lagrange's projection goes a stage further, its constants being exhausted at the second degree term in I. Now both these projections are special cases of one in which

$$y + ix = A \tan^n \frac{f - \gamma}{2},$$

and
$$\tan \frac{f}{2} = e^{m(\psi + a + iD)},$$

for if we put $m = 1$, $a = 0$ we have the conical orthomorphic with pole at the point γ, and if $n = 1$, $\gamma = \frac{\pi}{2}$, we have the Lagrange projection. The above equations introduce two more constants and so will take us a stage further.

Geometrically the projection is constructed in a manner similar to that for the Lagrange on p. 63, except that, with that figure, $NC = \gamma$ instead of $\frac{\pi}{2}$, and instead of taking simply $\tan \frac{CP}{2}$ we take the nth power of that quantity, and instead of ϕ, $n\phi$. To expand the co-ordinates put $f = F - u$, and $F - \gamma = \beta$; u can then be expanded in powers of l, the difference of latitude $\chi - \mu$, measured from the central parallel χ, and D, the longitude.

Suppose first of all $D = 0$.

Then we find

$$u_0 = \frac{\rho_0 \, m \sin F}{\xi_0} \left[l + \rho_0 \frac{(m \cos F - \cos \chi) \, l^2}{2\xi_0} \right.$$
$$+ \frac{\rho_0 l^3}{6\xi_0{}^2} (2\rho_0 \cos^2 \chi + \xi_0 \sin \chi + m^2 \rho_0 \cos 2F$$
$$\left. - 3m\rho_0 \cos \chi \cos F) \dots \right],$$

where u_0 has been taken to vanish at the centre, a condition which gives

$$e^{m(\psi_0 + a)} = \frac{e^{m\psi_0}}{k} = \tan \frac{F}{2}.$$

Next we have

$$\frac{du}{dD} = - mi \sin (F - u),$$

and thus when $D = 0$

$$\left(\frac{du}{dD}\right)_0 = - mi \sin F \left(1 - u_0 \cot F - \frac{u_0^2}{2} \cdots\right).$$

Continuing in this way we find

$$u = \frac{m \sin F}{\xi_0} \left[\rho_0 l - \frac{(m\rho_0^2 \cos F - \rho_0^2 \cos \chi) l^2}{2\xi_0} \right.$$

$$+ \frac{\rho_0^2 l^3}{6\xi_0^2} (2\rho_0 \cos^2\chi + \xi_0 \sin \chi + m\rho_0 \cos 2F - 3m\rho \cos \chi \cos F)$$

$$- iD\xi_0 + mi \cos F\rho_0 lD - mi \rho_0^2 \frac{l^2 D}{2} (m \cos 2F - \cos \chi \cos F)$$

$$\left. + m \cos F\xi_0 \frac{D^2}{2} - m^2 \cos 2F\rho_0 \frac{lD^2}{2} + m^2 i \cos 2F\xi_0 \frac{D^3}{6} \cdots \right].$$

Now

$$y + ix = A \tan^n \frac{\beta - u}{2}$$

$$= A \tan^n \frac{\beta}{2} \left(1 - \frac{K_1 (u_1\beta)}{\sin \beta} u + \frac{K_2 (u_1\beta)}{\sin^2 \beta} \frac{u^2}{2} \cdots\right).$$

Let the origin be moved to the centre, the direction of y reversed, and put in the condition that the scale is true there, and a minimum. These give

$$A \tan^n \frac{\beta}{2} nm \sin F = \xi_0 \sin \beta,$$

and

$$(n - \cos \beta) m \sin F = (\cos \chi - m \cos F) \sin \beta,$$

which, if n be put equal to unity and F to $\frac{\pi}{2} + \beta$, leads to the relation

$$\frac{e^{m\psi_0}}{k} = \frac{m - \cos \chi}{m + \cos \chi},$$

already found for Lagrange's projection. We then find

$$y = \rho_0 l + \xi_0 \cos \chi \frac{D^2}{2} + \frac{\rho_0^2 l^3}{6\xi_0^2}(\xi_0 \sin \chi + m\rho_0 Z)$$
$$- \frac{\rho_0 l D^2}{2}(mZ + \cos^2 \chi),$$

$$x = \xi_0 D - \rho_0 \cos \chi l D + \frac{\rho_0^2 l^2 DmZ}{2\xi_0} - \frac{\xi_0 D^3}{6}(mZ + \cos^2 \chi),$$

where

$$Z = (\cos \chi - m \cos F)(\cos F - \sin F \cot \beta).$$

These give for the scale error

$$I = \frac{\xi_0 \sin \chi D^2}{4\rho_0}\left(1 - \frac{\xi_0 \sin \chi + 2m\rho_0 Z}{\xi_0 \sin \chi}\right)$$
$$+ \frac{\rho_0 \sin \chi l^2}{4\xi_0}\left(1 + \frac{\xi_0 \sin \chi + 2m\rho_0 Z}{\xi_0 \sin \chi}\right)$$
$$= \frac{x^2}{4\rho_0 \nu_0}\left(1 - \frac{\xi_0 \sin \chi + 2m\rho_0 Z}{\xi_0 \sin \chi}\right)$$
$$+ \frac{y^2}{4\rho_0 \nu_0}\left(1 + \frac{\xi_0 \sin \chi + 2m\rho_0 Z}{\xi_0 \sin \chi}\right).$$

We must therefore take

$$2m\rho_0 Z = -(1+p)\xi_0 \sin \chi,$$

or $\quad \cot \beta = \cot F + \dfrac{(1+p)\xi_0 \sin \chi}{2m\rho_0 \sin F(\cos \chi - m \cos F)},$

where

$$p = \frac{r^2 - 1}{r^2 + 1} \text{ or } \frac{r^4 - 1}{r^4 + 1},$$

according to which of the two methods of determining it we use.

Now let us look at some special cases. In the conical orthomorphic for the sphere $m = 1$, $\alpha = 0$ which makes $F = \chi$, $u = \cos \beta = \cos(\chi - \gamma)$ and the expansions, as far as we have taken them, are independent of β and therefore of γ, and so will not serve to determine them. In the stereographic however β vanishes, $\chi = \gamma$, $p = 0$ and not -1 as in the former case, and the expansions are as on p. 136, while those for the other are the same as for the normal case of one standard parallel on p. 48.

In the conical orthomorphic for the spheroid we find

$$\cos F = \cos \chi \, (1 - \epsilon^2 \sin^2 \chi)$$

and

$$\cot \gamma = \cot \chi \left(1 + \epsilon^2 \frac{1 - p}{1 + p} \right).$$

In the Lagrange

$$F = \frac{\pi}{2} + \beta, \quad n = 1,$$

and we find

$$m \tan \frac{\beta}{2} = - \cos \chi,$$

and finally

$$m^2 = 1 + \epsilon^2 \sin^4 \chi + p \sin^2 \chi \, (1 + \epsilon^2 \sin^2 \chi),$$

whereas Lagrange chose m to make the second differential coefficient of the scale vanish, which means that p is arbitrarily chosen as unity, giving

$$m^2 = 1 + \sin^2 \chi + 2\epsilon^2 \sin^4 \chi,$$

as found on p. 64. The general expansions now become

$$y = \rho_0 l + \xi_0 \cos \chi \frac{D^2}{2} + \frac{\rho_0^2}{12\xi_0} \sin \chi \, (1 + p) \, l^3$$

$$+ \left(\xi_0 \sin \chi \frac{(1 + p)}{2} - \rho_0 \cos^2 \chi \right) \frac{lD^2}{2} \ldots,$$

$$x = \xi_0 D - \rho_0 \cos \chi l D + \rho_0 \, (1 + p) \sin \chi \frac{l^2 D}{4}$$

$$- \xi_0 \left(\cos^2 \chi - \frac{(1 + p) \, \xi_0 \sin \chi}{2\rho_0} \right) \frac{D^3}{6} \ldots.$$

For the calculation of β and F in the general case we must apply the conditions of minimum error to the succeeding terms; and this method might be extended further by taking x and y calculated as above, putting $y + ix = v$, and making a new projection

$$Y + iX = \tan^i \, (\delta + v),$$

thus introducing two more constants; and so on. But for most purposes the above formulae will be sufficient.

INDEX

Printed in the United States
By Bookmasters